Marketing in Foodservice Operations

David K. Hayes
Jack D. Ninemeier

SENIOR EDITORIAL DIRECTOR	Justin Jeffryes
EXECUTIVE EDITOR	Todd Green
EDITORIAL ASSISTANT	Kelly Gomez
SENIOR MANAGING EDITOR	Judy Howarth
PRODUCTION EDITOR	Mahalakshmi Babu
COVER PHOTO CREDIT	© ArtistGNDphotography/Getty Images

This book was set in 9.5/12.5 STIX Two Text by Straive™.
Published by John Wiley & Sons, Inc., Hoboken, New Jersey.
Published simultaneously in Canada.
This book is printed on acid-free paper.

Founded in 1807, John Wiley & Sons, Inc. has been a valued source of knowledge and understanding for more than 200 years, helping people around the world meet their needs and fulfill their aspirations. Our company is built on a foundation of principles that include responsibility to the communities we serve and where we live and work. In 2008, we launched a Corporate Citizenship Initiative, a global effort to address the environmental, social, economic, and ethical challenges we face in our business. Among the issues we are addressing are carbon impact, paper specifications and procurement, ethical conduct within our business and among our vendors, and community and charitable support. For more information, please visit our website: www.wiley.com/go/citizenship.

ISBN: 978-1-394-20833-3 (PBK)

Library of Congress Cataloging-in-Publication Data

Names: Hayes, David K., author. | Ninemeier, Jack D., author.
Title: Marketing in foodservice operations / David K. Hayes, Jack D.
   Ninemeier.
Description: Hoboken, New Jersey : Wiley, [2024] | Includes index.
Identifiers: LCCN 2023023162 (print) | LCCN 2023023163 (ebook) | ISBN
   9781394208333 (paperback) | ISBN 9781394208371 (adobe pdf) | ISBN
   9781394208388 (epub)
Subjects: LCSH: Food service—Marketing.
Classification: LCC TX911.3.M3 H38 2024 (print) | LCC TX911.3.M3 (ebook)
   | DDC 647.940688—dc23/eng/20230524
LC record available at https://lccn.loc.gov/2023023162
LC ebook record available at https://lccn.loc.gov/2023023163

The inside back cover will contain printing identification and country of origin if omitted from this page. In addition, if the ISBN on the back cover differs from the ISBN on this page, the one on the back cover is correct.

SKY10057538_101423

Contents

Preface *ix*
Acknowledgments *xiii*
Dedication *xv*

1 **Marketing for Foodservice Operations** *1*
   The Importance of Marketing *2*
      Who *3*
      What *4*
      Where *4*
      When *5*
      How *5*
   The 4 Ps of Marketing *5*
      Product *6*
      Place *7*
      Price *8*
      Promotion *9*
   Operators' Challenges in Marketing Products *11*
      Quality *11*
      Quantity *12*
      Delivery Method *12*
      Value Perception *13*
   Operators' Challenges in Marketing Services *14*
      Intangibility *15*
      Inseparability *15*
      Consistency *16*
      Limited Capacity *17*

2 Target Market Identification *21*

The Importance of Target Market Identification *22*

Why Target Markets Must Be Identified *23*

How Target Markets Are Identified *24*

Identifying the Needs of a Target Market *29*

Internal Factors Affecting Target Markets *33*

Physical Facilities *33*

Staff Capabilities *35*

External Factors Affecting Target Markets *38*

The Economic Environment *38*

Legal Requirements *39*

Vendor and Product Availability *40*

The Competitive Environment *40*

The Social Environment *42*

Technology *43*

3 Creating the Marketing Message *48*

The Marketing Message and Importance of Branding *49*

The Marketing Message *50*

Branding Defined *50*

Developing a Unique Brand *51*

Product-Focused Branding *54*

Service-Focused Branding *54*

Factors Affecting Product-Related Messages *55*

Product Quality *56*

Preparation Method *58*

Serving Size *58*

Price *59*

Convenience *61*

Location *62*

Service Offering Messages by Type of Operation *64*

Non-Commercial Operations *65*

Quick Service Restaurants (QSRs) *66*

Fast Casual Restaurants *66*

Casual Service Restaurants *67*

Fine Dining/Upscale Restaurants *68*

Ghost Operations *69*

4 Delivering the Marketing Message *72*

Assessing Marketing Channels *73*

Marketing Channel Options *73*

Assessing New Communication Channels *81*

Understanding Buyer Behavior *83*
   The Four Buyer Types *83*
   Buyer Behavior *86*
Operator Responsibilities in Target Market Messaging *88*
   Legal Responsibilities *89*
   Ethical Responsibilities *90*
   Social Responsibilities *91*

5 Creating the Marketing Plan *95*
The Need for a Written Marketing Plan *96*
   Advantages of Developing a Formal Marketing Plan *98*
   Challenges in Marketing Plan Development *99*
Creating the Marketing Plan *101*
   Identification of Marketing Strategies *104*
   Identification of Marketing Tactics *106*
   Creation of Marketing Plan Cost Estimates *108*
Implementing the Marketing Plan *109*
   Determining What Will Be Done *109*
   Determining When It Will Be Done *110*
   Determining Who Will Do It *110*
   Importance of Staff Efforts to Implement Marketing Plans *111*
Evaluation of Marketing Plan Results *115*

6 The Importance of Price in Marketing Efforts *119*
Pricing for Profits *120*
   The Importance of Pricing *121*
   The Operator's View of Price *121*
   The Guest's View of Price *122*
Factors Affecting Menu Pricing *124*
   Economic Conditions *124*
   Local Competition *125*
   Level of Service *125*
   Type of Guest *126*
   Product Quality *126*
   Portion Size *127*
   Delivery Method *128*
   Meal Period *129*
   Location *129*
   Bundling *130*
Methods of Food and Beverage Pricing *131*
   Cost-Based Pricing *132*
   Contribution Margin-Based Pricing *135*

Evaluation of Pricing Efforts   *137*
   Menu Engineering   *137*
   Menu Modifications   *141*

7   The Menu as a Marketing Tool   *145*
The Importance of the Menu   *146*
   The Menu as a Marketing Tool   *147*
   Types of Menus   *147*
   The Menu Development Team   *149*
   Legal Aspects of Menu Design   *149*
Creating the Food Menu   *151*
   Identifying Menu Categories   *152*
   Selecting Individual Menu Items   *153*
   Writing Menu Copy   *154*
   Key Factors in Successful Food Menu Design   *155*
Creating the Beverage Menu   *157*
   Beer Menus   *157*
   Wine Lists   *158*
   Spirit Menus   *159*
Digital Display Menus   *162*
   On-site Digital Menus   *162*
   On-user Device Digital Menus   *163*
Special Concerns for Off-Premise Menus   *165*
   Choosing Takeaway Menu Items   *166*
   Packaging Takeaway Menu Items   *166*
   Holding Takeaway Menu Items for Pickup or Delivery   *168*

8   Importance of the Foodservice Marketing Mix   *172*
Advertising   *173*
   The Purpose of Advertising   *174*
   Developing the Advertising Message   *176*
   Determining the Best Message Delivery Channel   *178*
Personal Selling   *179*
   The Goals of In-person Selling   *180*
   On-Premise Personal Selling   *181*
   Off-Premise Personal Selling   *182*
Promotions   *183*
   The Purpose of Promotions   *183*
   Promotions for All Customers   *184*
   Promotions for Loyal Customers   *185*

Publicity   *187*
  Initiating Positive Publicity   *187*
  Addressing Negative Publicity   *189*
Public Relations   *191*
  Media Affiliations and Successful Public Relations   *191*
  Community Relations and Successful Public Relations   *192*
  Employee Relations and Successful Public Relations   *193*

9   Web-Based Marketing on Proprietary Sites   *197*
Choosing the Primary Domain Name   *198*
Choosing the Website Host   *201*
The Proprietary Website   *204*
  Website Design   *206*
  Website Content   *208*
  Website Traffic Tracking   *211*
Search Engine Optimization   *212*
  How Search Engines Work   *213*
  Choosing Keywords and Phrases   *214*
Social Media Sites   *216*
  The Importance of Social Media   *216*
  Posting Content on Social Media Sites   *217*

10   Web-Based Marketing on Third-Party Sites and Apps   *222*
The Importance of Local Linkage   *223*
Marketing on Third-Party-Operated Websites   *226*
Partnering with Third-Party Delivery Apps   *231*
  Advantages of Third-Party Delivery App Partnerships   *232*
  Disadvantages of Third-Party Delivery App Partnerships   *234*
Self-Delivery Options   *239*

11   Marketing Management on User-Generated Content Sites   *245*
The Importance of User-Generated Content (UGC) Sites   *246*
Advertising on User-Generated Content (UGC) Sites   *247*
Popularity of User-Generated Content (UGC) Sites Featuring Guest
Reviews   *249*
  Historical Perspective   *250*
  Motivation of Reviewers   *252*
Improving Scores on User-Generated Content (UGC) Review Sites   *255*
  Increasing the Number of User Reviews   *256*
  Increasing the Scores on User Reviews   *259*
Responding to Negative User Reviews   *261*

12 Assessment of Marketing Efforts   *266*

The Importance of Evaluating Marketing Efforts   *267*
   Assessment of Marketing Strategies   *269*
   Assessment of Marketing Tactics   *270*
Marketing Evaluation Tools   *273*
   Tools for Financial Evaluations   *274*
   Tools for Nonfinancial Evaluation   *281*
Preparing Next-Period Marketing Plans   *284*
   SWOT Analysis   *284*
   Setting Next-Period Strategies and Goals   *288*
Summary   *289*

Glossary   *G-1*
Index   *I-1*

Preface

Every foodservice operation must offer high-quality menu items and provide excellent service to its guests, but these factors are still not enough to ensure the operation's long-term success. All operators must also be able to communicate their product and service offerings to their current and potential customers if they hope to grow their businesses.

Effective marketing is the process foodservice operators utilize to do just that. Marketing is the process used by businesses to inform potential customers about what it is offering for sale, and most importantly, why customers should buy from them.

The major goal of effective foodservice marketing is to obtain and retain a growing base of satisfied customers to help ensure a business can reach its financial goals. This book was written to address the most important information and marketing tools foodservice operators need to be successful in marketing.

The successful marketing of foodservice operations is very different today than it has been in the recent past. Traditional marketing and advertising methods such as newspaper and Yellow Page ads, brochures, and printed flyers have been replaced in large part by digital marketing efforts. In fact, today every food service operation, regardless of its size, must develop and maintain an effective online presence if it hopes to succeed in an increasingly competitive food service environment.

This book is unique in that it addresses traditional as well as the large and ever-growing impact of online communication in the marketing efforts of successful foodservice operators. Regardless of the segment in which they do business, in this book, foodservice operators (and those who aspire to be effective foodservice operators!) will discover what they must know and do to achieve marketing success.

Among the many key marketing topics addressed in this book are:

✓ The importance of effective marketing
✓ How to identify a target market
✓ How to create a marketing plan
✓ How menu prices impact marketing

✓ How to use the menu as a marketing tool
✓ The importance of a social media presence
✓ How to market using an operation's proprietary website
✓ How to optimize search engine results
✓ How to market with third-party content-controlled sites and apps
✓ How to market on popular user-generated content sites
✓ How to improve rating scores on user review sites
✓ How to respond to negative reviews posted online
✓ How to assess an operation's overall marketing efforts

Most importantly, all of these marketing topics, and more, are addressed in ways specific to the foodservice industry.

Marketing in the foodservice industry is unique because foodservice operators must promote outstanding menu items and exceptional service. This can be a significant challenge regardless of whether the business being marketed is a food truck, coffee kiosk, ghost restaurant, quick service restaurant, or full-service operation.

Readers will quickly find that the content of this book is essential to the profitable management of their own operations, and they will also find that the information in each chapter has been carefully selected to be easy to read, easy to understand, and easy to apply.

Book Features

In addition to the essential foodservice operations marketing information it contains, special features were carefully crafted to make this learning tool powerful but still easy to use. These features are:

1) **What You Will Learn** To begin each chapter, this very short conceptual bullet list summarizes key issues readers will know and understand when they complete the chapter.

2) **Operator's Brief** This chapter opening overview states what information will be addressed in the chapter and why it is important. This element provides readers with a broad summation of all important issues addressed in the chapter.

3) **Chapter Outline** This two-level outline feature makes it quick and easy for readers to find needed information within the body of the chapter.

4) **Key Terms** Professionals in the foodservice industry often use very special terms with very specific meanings. This feature defines important (key) terms so readers will understand and be able to speak a common language as they discuss issues with their colleagues in the foodservice industry. These key terms are also listed at the end of each chapter in the order in which they initially appeared.

5) **Find Out More** In a number of key areas, readers may want to know more detailed information about a specific topic or issue. This useful book feature gives readers specific instructions on how to conduct an Internet search to access that information and why it will be of importance to them.

6) **Technology at Work** Advancements in technology play an increasingly important role in many aspects of foodservice operations. This feature was developed to direct readers to specific technology-related Internet sites that will allow them to see how advancements in technology can assist them in reaching their operating goals.

7) **"What Would You Do?"** These "mini" case studies located in every chapter of the book take the information presented in the chapter and use it to create a true-to-life foodservice industry scenario. They then ask the reader to think about their own response to that scenario (i.e. *What Would You Do?*).

   This element was developed to help heighten a reader's interest and to plainly demonstrate how the information presented in the book relates directly to the practical situations and challenges foodservice operators face in their daily activities.

8) **Operator's 10-Point Tactics for Success Checklist** Each chapter concludes with a checklist of tactics that can be undertaken by readers to improve their operations and/or personal knowledge. For example, in a chapter of the book related to marketing with social media, one point in that chapter's 10-Point Tactics for Success Checklist is:

   Operator understands the value of regularly posting and updating content on their chosen social media sites.

Instructional Resources

This book has been developed to include learning resources for instructors and students.

To Instructors

To help instructors (and corporate trainers!) effectively manage their time and enhance student-learning opportunities, the following resources are available on the instructor companion website at www.Wiley.com/go/hayes/marketingfoodservice.

✓ Instructor's Manual that includes author commentary for "What Would You Do" mini case-study questions.

✓ PowerPoint slides for instructional use in emphasizing key concepts within each chapter.

✓ A 100-item Test Bank consisting of multiple-choice exam questions, their answers, and the location within the book from which the question was obtained. The test bank is available as a print document and as a Respondus computerized test bank. Note: **Respondus** is an easy-to-use software program for creating and managing exams that can be printed to paper or published directly to Blackboard, WebCT, Desire2Learn, eCollege, ANGEL, and other e-Learning systems.

To Students

Learning about foodservice marketing will be fun. That's a promise from the authors to you. It is an easy promise to make and keep because working in the foodservice industry is fun. And it is challenging. However, if you work hard and do your best, you will find that you can master all of the important information in this book.

When you do, you will have gained invaluable knowledge that will enhance your skills and help advance your own hospitality career. To help you learn the information, in this book, online access to over 220 PowerPoint slides is available to you. These easy-to-read tools are excellent study aids and can help you when taking notes in class.

Acknowledgments

Marketing in Foodservice Operations has been designed to be the most up-to-date, comprehensive, technically accurate, and reader-friendly learning tool available to those who want to know how to increase profits by effectively marketing their foodservice operations.

The authors thank Catriona King of Wiley for initially working with us to develop the idea for a series of practical books that would help foodservice operators of all sizes more effectively manage their businesses. She was essential in helping conceptualize the need for this book as well as all of the other books in this 5-book Foodservice Operations series. The five titles in the series are:

✓ *Marketing in Foodservice Operations*
✓ *Cost Control in Foodservice Operations*
✓ *Accounting and Financial Management in Foodservice Operations*
✓ *Managing Employees in Foodservice Operations*
✓ *Successful Management in Foodservice Operations*

We would also like to thank the external reviewers who gave so freely of their time as they provided critical industry and academic input on this series. To our reviewers, Dr. Lea Dopson, Gene Monteagudo, Isabelle Elias, and Peggy Richards Hayes, we are most grateful for your comments, guidance, and insight. Also, thanks to Michael T. Kavanagh, who was a technological friend indeed, when we were most in need!

Books such as this require the efforts of many talented specialists in the publishing field. The authors were extremely fortunate to have Todd Green, Judy Howarth, and Kelly Gomez at Wiley as our publication team. Their efforts went far in helping the authors present the book's material in its best and clearest possible form.

Finally, the authors would like to thank the many students and industry professionals with whom we have interacted over the years. We sincerely hope this book allows us to give back to them as much as they have given to us.

David K. Hayes, Ph.D.
Jack D. Ninemeier, Ph.D.

Dedication

The authors are delighted to have the opportunity to dedicate this book, and this entire Foodservice Operations series, to two outstanding and unique individuals.

Brother Herman Zaccarelli

Brother Herman E. Zaccarelli, C.S.C., passed away in 2022 at the Holy Cross House in Notre Dame, Indiana. His professional work included many projects for the hospitality industry, and he published several books and hundreds of articles for numerous trade publications over many years. Among numerous accomplishments, Herman founded Purdue University's Restaurant, Hotel, and Institutional Management Institute in 1976. Later, he served as Director of Business and Entrepreneurial Management at St. Mary's University in Winona, Minnesota.

A lifelong learner, at the age of 68, Brother Herman retired to Florida where he earned a Bachelor's degree in Educational Administration and a Master's degree in Institutional Management.

Herman's ideas and concepts have been widely adopted in the hospitality industry, and he assisted many young educators including the authors of this book series. He will be remembered as a colleague with creative ideas who provided significant assistance to those studying and managing in the hospitality industry. Herman was especially helpful in discovering and addressing learning opportunities for Spanish-speaking students, educators, and managers throughout the United States and around the world.

Dr. Lea R. Dopson

A lifelong friend, advisor, and colleague, as well as an outstanding author herself; at the time of her untimely passing, Lea served as President of the International Council on Hotel, Restaurant, and Institutional Education (ICHRIE) and Dean of the prestigious Collins College of Hospitality Management (Cal Poly Pomona).

Lea was a dedicated hospitality professional and a fierce advocate for hospitality students at all levels. Those who knew her were continually in awe of her intelligence and humility.

It was especially fitting that Lea was named as a recipient of the H.B. Meek Award. That award is named after the individual who started the very first hospitality program in the United States (at Cornell University). Selected by the recipient's peers, it goes not to the most outstanding academic professional working in the United States but to the most outstanding academic professional in the entire world. That was Lea.

While she is dearly missed, her inspiration goes on everlastingly in the works of the authors.

1

Marketing for Foodservice Operations

> **What You Will Learn**
>
> 1) The Importance of Marketing
> 2) The 4 Ps of Marketing
> 3) Operators' Challenges in Marketing Products
> 4) Operators' Challenges in Marketing Services

Operator's Brief

In this chapter, you will learn that every foodservice operation must offer high-quality menu items and provide excellent service to its guests, but these factors are not enough to ensure the operation's long-term success.

All foodservice operators must also be able to communicate their product and service offerings to their current and potential customers if they hope to grow their businesses. Marketing is the process operators utilize to do just that.

One good way to view the marketing process is to consider the 4 Ps of marketing: product, place, price, and promotion. Products are the menu items and services an operation offers for sale. Place refers to the physical location where the operator will deliver its products to guests. Price is the amount the operation will charge its guests for what it sells. Promotion refers to the many communication methods and tools operators use to deliver their important marketing messages.

Marketing effectively presents many challenges to foodservice operators as they seek to promote their products and services. Specific challenges when promoting foodservice products include variations in product quality, product quantity, methods of product delivery, and the value perceptions of guests.

(Continued)

Food and beverage operations are part of the service industry, and operators of these businesses face additional marketing challenges. For example, while all operators want to provide excellent service to guests, service quality is actually affected by a number of unique characteristics of service. These characteristics include intangibility, inseparability, consistency, and limited capacity. In this chapter, you will learn about all of these challenges and what successful operators must do to recognize and address them.

CHAPTER OUTLINE

The Importance of Marketing
    Who
    What
    Where
    When
    How
The 4 Ps of Marketing
    Product
    Place
    Price
    Promotion
Operators' Challenges in Marketing Products
    Quality
    Quantity
    Delivery Method
    Value Perception
Operators' Challenges in Marketing Services
    Intangibility
    Inseparability
    Consistency
    Limited Capacity

The Importance of Marketing

All foodservice operations must effectively serve their customers. Whether an operation's customers are called guests, students, patients, or with any other titles, satisfying those being served is essential for the continued operation of any successful foodservice. In fact, foodservice operators should actually consider their main job to be that of adding new customers while retaining their current customers.

The Importance of Marketing | 3

To ensure a steady flow of customers, foodservice operators must understand **marketing**. Marketing includes all the ways a foodservice operation communicates with its current and potential guests.

Marketing is how a business informs potential customers about what it is offering for sale, and most importantly, why customers should buy from them. The major goal, or purpose, of effective marketing is to obtain and retain a growing base of satisfied customers.

Key Term

Marketing: The varied activities and methods used to communicate a business' product and service offerings to its current and potential customers.

Effective marketing helps build a positive relationship between a business and its customers. Marketing is different from selling. It is not possible to sell a customer something unless a company's marketing efforts have brought the customer to the point where he or she is willing to buy.

Before foodservice operations can explain "why" guests should buy from them, they must first communicate some basic and important information about themselves to their potential customers. This includes:

✓ Who
✓ What
✓ Where
✓ When
✓ How

Who

Today's consumers seeking food and beverage services may have virtually hundreds of operations to choose from. Therefore, the name of an operation is important because it is the first step in establishing a communication link with potential guests.

Some food operations choose clever names that are easy for potential guests to remember. Others may incorporate the menu items they will serve into the names of their operations, while still others may choose names that simply identify the owner of the operation.

Regardless of the naming approach taken, potential customers must be able to easily identify and remember an operation by its name if they are to learn more about it and become actual customers.

What

Because of the great variety of foodservice operations open today, owners of foodservice operations must communicate to their potential guests exactly what they offer for sale. When an operation's name carries this information (for example, Tim's Pizza or Taco Time), guests may easily understand both who the operation is and the types of products it offers.

If, however, an operation's name does not carry any specific product and service offerings information (for example, Cindy's Restaurant, or the one-word name, "Carson's"), then operators must take additional steps to let potential guests know exactly what types of menu items the operation offers for sale.

In addition to on-premises dining, some operations offer services such as offsite catering or special onsite banquet rooms that may be reserved. When available, this type of information must also be communicated to guests.

Key Term

Ghost kitchen: A foodservice operation that provides no on-premises dine-in services and that prepares all of its menu items only for pick-up or delivery to guests. Also known as a delivery-only restaurant, shadow kitchen, cloud kitchen, or virtual kitchen.

Where

After guests know who they will buy from, and what they may buy, these guests must know where they will go to buy it. One way to communicate the location of a restaurant to guests is through visible exterior signage. That can work well in some cases; however, many potential guests who know the name of an operation may want the address and driving instructions to it using the information posted online and/or in the operation's printed marketing materials.

In some cases, for example, with food trucks, an operation's physical location is not fixed and the location of the truck changes regularly so its changing location must be communicated to potential guests on a daily basis.

In still other cases, for example, in operations that produce their food in a **ghost kitchen**, a physical location must be provided to guests or **third-party delivery** personnel who will take the food to the point of consumption. Note: The advantages and disadvantages of third-party delivery partnerships will be addressed in detail in Chapter 10.

Key Term

Third-party delivery: A smartphone or computer application that creates a marketplace that customers can search to browse restaurant menus, place orders, and have them delivered to a location of the customer's choosing. In nearly all cases, the guest orders are delivered by independent contractors who have been retained by the company operating the third-party delivery app.

When

Information about a foodservice operations' operating hours are often important to communicate to customers because most restaurants are not open 24 hours a day and 7 days a week. Thus, the days of the week an operation is open, and closed, as well as the time they begin their service day and end their service day, is critically important information for potential guests.

How

In many cases, guests choosing to buy products from a foodservice operation may have several choices for placing their orders and making payments for the products purchased. While some restaurants feature exclusively on-premises dine-in service, many operations also offer drive-through, takeaway, or third-party delivery choices for their guests. When alternative choices are available, all of those options must be effectively communicated.

Traditionally, foodservice guests have paid for their purchases on-site at the time they placed their orders or after finishing their meals. Payment forms consisted primarily of cash or credit cards. In increasing numbers of operations, however, guests may have additional payment options available to them by using their smart devices and various payment apps. In all cases, effective marketing includes informing guests about the different ways they can purchase the menu items they select.

The 4 Ps of Marketing

With so much critical information to be shared with guests, many marketing experts find it convenient to focus on the **4 Ps of marketing** to increase the likelihood of their success when advertising their products and services.

The 4 Ps of marketing are illustrated in Figure 1.1.

The actual use of the 4 Ps of marketing is sometimes referred to as an operation's **marketing mix**, because different foodservice operators emphasize each of these four components in different ways as they market their businesses. An operation's specific marketing mix choices are so important they will be addressed in detail in Chapter 8.

Key Term

4 Ps of Marketing: A means of categorizing a business' marketing strategy on the basis of the products sold, places where they are sold, the prices at which they are sold, and the promotional efforts used to sell them.

Key Term

Marketing mix: The specific ways a business utilizes the 4 Ps of marketing to communicate with its potential customers.

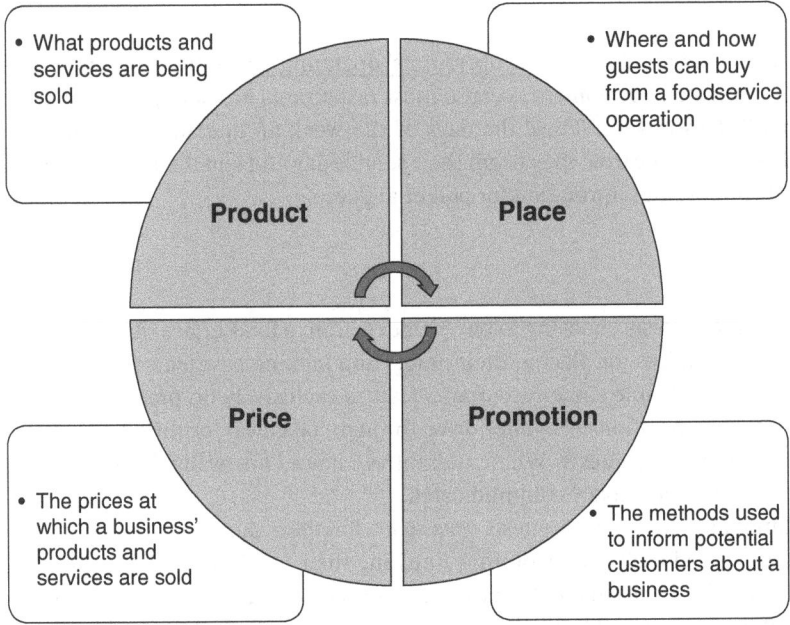

- What products and services are being sold

Product

- Where and how guests can buy from a foodservice operation

Place

Price

Promotion

- The prices at which a business' products and services are sold

- The methods used to inform potential customers about a business

Figure 1.1 The 4 Ps of Marketing

Product

In a foodservice operation, the specific menu items sold, and the ways they are sold to guests, are the operation's products. Most foodservice operators spend a lot of time thinking about and analyzing the products they sell because they know they must consistently satisfy their guests' product needs. Consequently, most foodservice operators are very proud of the menu items they offer and have spent significant time developing the recipes and cooking methods used to create their products.

Products offered for sale are, of course, important to guests, but the products sold by foodservice operators can vary widely from one operation to the next. For example, it is well-known that hamburgers are the single most popular menu item sold in the United States. In some operations, the production of a quality hamburger may involve cooking a 2-ounce (purchase weight) beef patty, and then placing it on a bun with ketchup, mustard, and a single pickle slice. The hamburger may then be wrapped and, when sold, it may be placed in a paper sack for delivery to a customer in the operation's drive-through lane.

In a different operation, a hamburger may be made by cooking to medium rare a 6-ounce (purchase weight) patty of extremely high-quality beef. The patty may then be topped with cheddar cheese, leaf lettuce, and a slice of red onion, and served on a brioche bun. This product may then be picked up in the kitchen by a waitstaff member who will deliver the hamburger to the guest's table.

In this example, both operations serve a "hamburger," but the products received by their guests will be very different. The prices paid by the guests will likely be very different as well. In foodservice operations, a menu item may vary by portion size, method of preparation, speed of delivery, and packaging method, to name but a few possible variations.

It is important to understand that just because a menu item is inexpensive does not mean it will be of low quality. All foodservice operations must consistently strive to serve their menu items in a way that optimizes the quality of the products ordered by their guests.

Place

Place refers to the physical location where products are available for sale or delivery. Location can be very important to the success of a foodservice operation. The location of an operation may include its indoor dining areas and outdoor patio areas, and it also includes the **curb appeal** of the operation's exterior, its lighting, furnishings, and exterior signage. Even the uniforms worn by staff and the background music being played within the operation may be considered a component of "Place."

When foodservice operations market "Place" as a key feature when appealing to their potential guests, they must consider the entire guest experience. That includes paying special attention to décor, the size of the type on its menus, and the even the quality of cleanliness in its guest restrooms!

Key Term

Curb appeal: The general attractiveness of a foodservice operation when viewed from the outside by a potential customer.

As previously indicated, ghost restaurants do not have a physical location to welcome their guests. For these operations, "Place" may include the quality of the vehicles used to deliver menu items, the quality of uniforms worn by delivery drivers, and even the appearance of the drivers themselves.

One good way for operators to understand "Place" as a key feature in the marketing message is to see the operation from their customers' viewpoints. That is, what do customers see, hear, and smell as they arrive at a foodservice operation's entrance area?

Convenient, well-lighted parking areas, clean entrance and waiting areas, and tidy dining areas are important to all guests regardless of the menu items being served. It is also important to recognize that, while cleverly designed buildings and beautiful interiors may draw guests for initial visits, only the proper serving of quality food and beverage products will encourage those guests to return.

Find Out More

For many foodservice operations, background music is an important component of "Place." Music can be used to support or reinforce an operation's theme, or to create a positive "feel" that is consistent with the operation's overall ambience.

But background music is not free. All foodservice operators playing recorded music must pay a fee to a Professional Rights Organization (PRO) or to a music service that has paid the appropriate fees on their behalf, to be able to play the music legally. There are three main PROs in the United States. These are SESAC, ASCAP, and BMI. Each one represents a specific segment of the music industry.

To learn more about the legal requirements for playing recorded music in public spaces including food and beverage operations, enter "professional rights organization music" in your favorite search engine and view the results.

Price

Price is so important to the success of foodservice operations that it will be the single topic addressed in Chapter 6. In many cases, the price of a product or service directly influences its sales volume and, as a result, a foodservice operation's revenue and profits.

Experienced foodservice operators know that the prices charged for the menu items they sell can be affected by numerous things. These include product and labor costs, overhead costs, the pricing trends of competitors, and even, in the case of the sale of alcoholic beverages, government regulations.

From a marketing perspective, menu prices send a very clear message to potential guests about the value an operation provided in its products and services. When foodservice operators use higher quality ingredients, or give larger portion sizes, menu prices will reflect that. Similarly, for those foodservice operations located in highly desirable areas, for example, those with ocean-view dining rooms or locations in popular and trendy downtown areas, prices may reflect that location's desirability as well.

Proper pricing is important because, if an operation's prices are set too low, the operation may be popular with guests, but it may also have difficulty achieving its financial goals. Similarly, if prices are set too high, the operation may attract too few customers. This may also cause the operation to have trouble achieving its financial goals.

As a part of its marketing function, "Price" also includes how guests actually pay for their purchases. Historically, foodservice operations accepted cash, checks, and credit and debit cards as payment. Today, few operations accept personal checks but increasingly they offer their guests the opportunity to pay through payment apps such as Apple Pay, Google Pay, or Samsung Pay. When guests are offered these payment options, that price-related information must also be properly communicated to them.

Promotion

Promotion is the portion of an operation's marketing efforts that seeks to communicate (speak) directly with customers. Key components of promotion include:

Advertising
Personal selling
Promotions
Publicity
Public relations

Each of these important activities will be addressed in detail in Chapter 8, but it is important to understand that advertising is an example of a food operation's direct messaging to guests. The key components of advertising include developing an effective advertising message and then selecting the best tools to deliver that message. Operators must also consider the frequency and cost of their message delivery.

Personal selling efforts in foodservice operations include both those efforts taken on-premises and those performed off-premises. On-premises selling efforts include signage about a featured item and recommendations by waitstaff and bartenders as they serve guests. Off-premises selling efforts are those undertaken by an operation's owners, managers, or sales staff to increase local business. Examples may include websites, social media activity, and participation in local community events.

"Promotion" is one of the 4 Ps of marketing, but the same term is also used to identify specific offers designed to appeal to all customers and, in many cases, to provide special rewards for loyal customers. For example, a foodservice operation may offer, as a special promotion, reduced pricing on a selected menu item during certain times of the day or week.

"Publicity" refers to communication about a business that has been placed in the media without the business paying for it directly. Publicity is important to a foodservice operation for two reasons.

The first is that the operation benefits from positive publicity since it reflects well on the business. For example, a foodservice operation may sponsor a youth sports team. If that sponsorship is reported in a local newspaper, or on television, radio, or a popular website, the publicity will likely be positive.

The second reason that publicity is important is that not all publicity is positive. For example, in many communities, sanitation inspection scores assigned by the local health department for foodservice operations are made public in the newspaper or on television.

If an operation's inspection score is not good, the result can be negative publicity for that operation. Similarly, a significant outbreak of a foodborne illness is likely to be widely reported in the local community. The publicity surrounding such an outbreak can have an extremely negative impact on an operation's reputation. For that reason, it is essential that all foodservice operations have in place an effective food handling and food safety training program for its **back-of-house staff**.

Key Term

Today, a large number of popular websites allow guests to post comments about their most recent visits to a foodservice operation. These reviews will no doubt be widely read by potential guests of the operation. While many posted comments may be positive, some will likely be negative. Proactively responding to these comments posted on reviewer sites is such an important topic that it will be addressed in detail in Chapter 11.

Back-of-house staff: The employees of a foodservice operation whose duties do not routinely put them in direct contact with guests.

Finally, foodservice operations must employ public relations to promote and manage a positive public image, which is how the operation is viewed by the general public. An effective marketing strategy for a foodservice operation involves identifying the right customers and then using all of these promotion tools to reach potential guests with a compelling sales message.

Find Out More

All foodservice operators must ensure that the products they sell are wholesome and safe to eat. No foodservice operation wants to be the source of a foodborne illness affecting guests. Therefore, it is important that every foodservice operator ensures their operation's back-of-house employees have been fully trained in safe food handling techniques.

Fortunately, the Internet offers a large number of food safety training tools that are inexpensive or even free to use. To find out more about these training tools, enter "food safety training videos for restaurants" in your favorite search engine and view the results.

Technology at Work
Understanding who their customers are and being able to easily communicate with them is an important tactic for all foodservice operations. Customer relations management (CRM) programs are designed to help them do just that.
CRM software helps foodservice operators maintain a centralized database of information on customers who have visited their operation or ordered online. Operations of all categories and sizes can use data such as this for marketing and sales promotions, email, or social media campaigns, and to create buyer loyalty programs.
To review specific ways CRM programs can assist foodservice operators, enter "restaurant customer service relations programs" in your favorite search engine and view the results.

Operators' Challenges in Marketing Products

Regardless of what they sell, all foodservice operators face unique challenges as they market their products. In fact, foodservice operators face four very unique challenges as they develop marketing messages for the products they sell. These product-related communication challenges are related to:

✓ Quality
✓ Quantity
✓ Delivery method
✓ Value perception

Quality

Every foodservice operator's menu items must be made with wholesome food: that which is safe to eat. However, the quality of ingredients used to make menu items can vary greatly, and meats, fruits, cheeses, coffee, wine, and other alcoholic beverage products provide just a few examples.

For example, the New York strip steak served in one steakhouse may be classified as USDA Prime, the highest quality ranking for beef that has been graded by the United States Department of Agriculture (USDA). A competitor of that operation may serve New York strip steaks that were graded as USDA Choice, the second highest quality ranking. In this example, both operations may advertise that they sell New York strip steaks, but the quality of the steaks is not the same, nor is the purchase price the operator must pay for the steaks.

As a result, part of the marketing message for the operation serving the higher quality Prime steaks must indicate that higher charges are necessary because guests are served the highest possible quality of beef.

Knowledgeable foodservice operators know that many ingredients used in the preparation of their menu items can vary widely in quality.

Quantity

There are no legal requirements dictating the quantity of a menu item that must be served in a foodservice operation. Thus, for example, a food operator might choose to serve a 3-, 4-, 5-, or 6-ounce hamburger patty. Similarly, an operation's "Fried Fish Basket" may include 1, 2, or 3 fried fish fillets and of varying sizes. Soft drinks may be sold as small, medium, or large, with the drink's container size determined by the operator.

Since portion sizes can vary so much, new guests frequently will have little idea about the actual quantity of food they are buying when they place a food order. Detailed menu descriptions and photographs of food items can help. However, guests must be alerted to the quantity of food or beverage they will receive for the amount they will pay if they are to be satisfied with their purchases.

Delivery Method

Different foodservice operators may offer identical products in identical portion sizes and yet, from their customers' perspectives, they are offering very different products. For example, one operator may serve a 5-ounce cup of coffee made with a particular brand of ground coffee, served in a paper cup with a plastic lid and delivered to the guest via the operation's drive-through window.

Another operator may offer the same 5-ounce cup of coffee, made with the same brand of coffee, but served on a saucer and in an elegant china cup in an exclusive upscale dining room. In this example, both product quality and quantity are identical, yet what each operation's guests will receive is very different.

In addition to the menu items they desire, many guests will want their menu selections served in a particular way, perhaps in a particular container, in a specific setting, or with a particular level of speed.

In many locations, guests have numerous alternatives including **quick service restaurants (QSRs)**, fast casual restaurants, or fine-dining operations. When guests have these alternatives, it can be a challenge to communicate how their product offerings are differentiated from those of other operators.

Value Perception

All foodservice operators must provide their guests with a good **value** for the products guests purchase. If they do not, the guests are unlikely to be satisfied and will not return. The challenge for foodservice operators in relation to value perception is to remember that value is determined by the buyer, not the seller, in any business transaction.

For example, a foodservice operator may sincerely believe that an item's quality, quantity, and delivery method will provide good value to guests if it is sold for $19.95. If, however, too few customers share that view of value, the operation will not consistently sell the menu item.

Key Term

Quick service restaurant (QSR): Foodservice operations that typically have limited menus, and most often include a counter at which customers can order and pick up their food. Most QSRs also have one or more drive-through lanes that allow customers to purchase menu items without leaving their vehicles. Menu prices in QSRs will be relatively low compared to other restaurant types.

Key Term

Value: The amount paid for a product or service compared to the buyer's view of what they receive in return.

Foodservice operators also must understand that providing good value to guests is not the same as providing low selling prices. Customers can perceive excellent levels of value in both high-priced foodservice operations and in those that charge more modest prices. Alternatively, guests may pay very low prices for menu items and still feel they have not received good value for the money they spent.

Experienced operators know that their customers assign the value of many foodservice products they purchase based on a number of factors in addition to quality, quantity, and method of delivery. This means that communicating value for items sold can often be extremely challenging.

To illustrate simply how the concept of value actually works, consider that, if a customer normally pays $19.99 for a medium pizza, a larger pizza for the *same* price represents an increase in value. A smaller pizza for the same price represents a *decrease* in value. Similarly, a reduction in price from $19.99 to $16.99 for a medium pizza represents an increased value.

In most cases, buyers make their value judgments regarding the wisdom of a purchase based on their own very personal assessment value. As a result, an important aspect of marketing foodservice products entails communicating to guests exactly "what they will get" and "why it is of value to them." This becomes even more challenging when the foodservice industry, as part of the hospitality industry, is classified as a **service industry**, not a manufacturing industry.

Foodservice operations do, of course, prepare (manufacture) food and beverage items that are then served to their guests. Clearly, the steaks and glasses of wine served to guests in a fine-dining restaurant are products. The presentation of the steaks and wine, however, is also clearly a service. Perhaps the reason foodservice operations are classified as members of the service industry is related to the importance of product quality *and* service quality to guests' perceptions of value.

The most successful marketers of foodservice products understand that in all segments of the hospitality industry, service quality is typically perceived by most guests as *more* critical to value delivered than product quality.

Key Term

Service industry: A business segment that primarily provides services to its customers.

Find Out More

Today's foodservice customers increasingly rely on websites posting consumer comments as they seek information about where to go and spend their "away-from-home" food dollars.

Sites such as Yelp, Facebook, OpenTable, and TripAdvisor are among the most popular of these. To see examples of reviews written by guests, enter the name(s) of one or more restaurants that are among your favorites and view the customer comments that have been posted about it.

Would the reviews you read encourage, or discourage, you from going there?

Operators' Challenges in Marketing Services

Foodservice operators sell products, and they deliver services. In most cases, it is easier for foodservice operators to clearly communicate information about the products they sell than it is to communicate information about the service levels they provide.

For example, a 16-ounce ribeye steak will always be larger than a 10-ounce ribeye. Priced properly, guests may perceive better value in the larger steak (such as when the cost per ounce is less for the larger steak). Similarly, French fries may be offered in small or large portion sizes. In both of these examples, the buyer can easily be told what they will be getting for the prices they will be charged.

When a foodservice operation advertises that it offers quality service, good service, or quick service, these concepts may not be so easy to communicate and deliver. Consider, for example, that an operation's service quality is affected by a number of unique characteristics of service. These characteristics are:

✓ Intangibility
✓ Inseparability
✓ Consistency
✓ Limited Capacity

Intangibility

Foodservice customers buy products, but they receive services. The sale of a service provides an intangible benefit to a foodservice guest. It is intangible because, unlike a physical product, a service cannot be seen, tasted, felt, heard, or smelled prior to its purchase. In most cases, a service is actually a performance rather than a product. Because they are intangible, operators face unique challenges in communicating the benefits of services that are offered to those who will buy them.

Guests buying from a foodservice operation must put their faith in the operator that the service provided will be of high quality. Foodservice operators must justify this trust by consistently delivering service at a level that meets, and even exceeds, their customers' expectations.

Inseparability

Inseparability refers to the tendency of foodservice guests to equate the quality of service they receive with the actual person who provides it. As a result, the *rude* waitstaff member will be perceived by guests as providing poor service, while the *cheerful* waitstaff member will be perceived as providing good service (even when the precise tasks performed by the two are similar or even identical).

It is for this reason that so many foodservice operators hire those who will directly serve guests for their attitudes, rather than their skills. These operators understand that the real job of a professional food server is to:

✓ Make guests feel welcome
✓ Make guests feel important
✓ Make guests feel special
✓ Make guests feel comfortable

✓ Show a genuine interest if guest expectations have not been met and corrective action must be taken

✓ Correct any service shortcomings promptly and with a positive attitude

Consistency

Consistency in service can be a challenge to provide because the quality of service often depends on the individual who supplies it. Inconsistency when providing services is usually much greater than when providing products. For example, the customer at a sports bar will usually find that the quality of several bottled beers of the same brand purchased while watching a game is identical; however, the skill level, appearance, amount of attention provided, and attitude of the beer's server can vary greatly, even during the same game.

In this example, these potential differences in product delivery can have a direct effect on the quality of the customer's beer purchase experience. Due to the importance of consistency when providing hospitality goods and services, it follows that foodservice operators should pay a great deal of attention to their operations' **front-of-house staff** training and standardization efforts.

Key Term

Front-of-house staff: The employees of a foodservice operation whose duties routinely put them in direct contact with guests.

What Would You Do? 1.1

"I can't get her attention," said Shingi. "I keep trying but she isn't looking over at us."

Shingi and her co-worker Annabella were having lunch at Oscar's. Oscar's was a restaurant popular for its soups and sandwiches served in a casual setting.

"I know you asked for a side of mayonnaise when you ordered your sandwich. She must've forgotten," said Annabella.

"I think your right," said Shingi, "I didn't notice it when she delivered our food. Now I'm not sure exactly what to do. She hasn't checked back with us."

"And your sandwich is getting cold," observed Annabella.

Assume you were the operator of the Oscar's. What do you think will be Shingi's assessment of the product quality of the sandwich she received during her visit? How can service quality directly impact a customer's view of product quality?

Find Out More
Service quality is important in every type of foodservice operation. As a result, service training for front-of-house employees is important in every operation. Fortunately, foodservice operators can find a variety of high-quality and inexpensive tools to help them in their guest service training efforts. To view some of these training tools, enter "restaurant server training programs" in your favorite search engine and view the results.

Limited Capacity

Nearly all foodservice operations face limitations in their capacity. In the front of the house (public areas), capacity may be limited by the number of seats in the dining room, or the number of drive-through lanes and windows available.

In the back of the house (employee-only areas), capacity may be limited by the number of workers available or the production capability of an operation's preparation equipment. As a result, it may be easier to deliver high-quality service during times of lower ("average") capacity than at times of full capacity. But foodservice operators seek to attract enough customers to operate at near full capacity because, unlike product inventory that can be held over from one day to the next, lost revenue from a seat that remained empty during a service period is lost forever.

Since most foodservice operations face limited capacity, it is especially important that their managers assign staff based on anticipated guest demand levels. Too many workers scheduled during slow times may cause an operation's labor-related costs to rise to unacceptable levels. However, too few workers available during busy times may cause an operation's service levels to suffer, even when its service workers are well-trained and are doing their very best.

The professional marketing of its products and services is essential to the success of every foodservice operation. The remaining chapters of this book address those things foodservice operators must know, and do, to market their operations successfully. When they are successful, they will find their efforts lead to the results desired by every foodservice operator:

✓ Increased recognition
✓ Increased customers
✓ Increased sales
✓ Increased profits

What Would You Do? 1.2

"I'm sorry sir," said Carmen, the dining room host at the Wagon Wheel restaurant. "I can put you on our waitlist, but I think it will be about 45 minutes before I can seat your party of four. We are always busy on Saturday nights."

"45 minutes?" replied the guest. "But I can see from here that the dining room has lots of empty tables right now."

"Yes sir," said Carmen. "But those tables have to be cleared and reset before we can seat additional guests. We only have one busser working tonight, and he's clearing tables as fast as he can."

"I have heard good things about this place, but I don't think I want to wait another 45 minutes. I think I'll just take our group somewhere else," replied the guest as he shook his head and left the host stand.

Assume you were the operator of the Wagon Wheel restaurant. What specific steps could you take to improve guest service on busy nights such as this one? Why would it be important for you to do so?

Key Terms

Marketing	Curb appeal	Value
Ghost kitchen	Back-of-house	Service industry
Third-party delivery	staff	Front-of-house staff
4 Ps of Marketing	Quick service	
Marketing mix	restaurant (QSR)	

Operator's 10-Point Tactics for Success Checklist

Evaluate your need for, and the current status of, each of the following operational tactics. For those tactics you think are important, but not yet in place, develop an action plan for its implementation including who will be responsible for the tactic's completion and the target date by which it should be completed.

Tactic	Don't Agree (Not Done)	Agree (Done)	Agree (Not Done)	Who Is Responsible?	Target Completion Date
				If Not Done	
1) Operator has carefully considered the relative importance of "Product" in the marketing mix for their own operation.	_____	_____	_____		

Tactic	Don't Agree (Not Done)	Agree (Done)	Agree (Not Done)	If Not Done	
				Who Is Responsible?	Target Completion Date
2) Operator has carefully considered the relative importance of "Place" in the marketing mix for their own operation.	____	____	____		
3) Operator has carefully considered the relative importance of "Price" in the marketing mix for their own operation.	____	____	____		
4) Operator has carefully considered the relative importance of "Promotion" in the marketing mix for their own operation.	____	____	____		
5) Operator has implemented an approved food safety training program to minimize the potential for negative publicity resulting from a food-borne illness outbreak.	____	____	____		
6) Operator has carefully considered potential quality- and quantity-related communication challenges in their own operation.	____	____	____		
7) Operator has carefully considered any potential delivery method and value-related communication challenges in their own operation.	____	____	____		
8) Operator has carefully considered the potential impact of service intangibility and inseparability in their own operation.	____	____	____		

(Continued)

				If Not Done	
Tactic	**Don't Agree (Not Done)**	**Agree (Done)**	**Agree (Not Done)**	**Who Is Responsible?**	**Target Completion Date**
9) Operator has carefully considered the potential impact of service consistency and limited capacity in their own operation.	_____	_____	_____		
10) Operator has implemented an effective guest service training program for front-of-house employees to help optimize service delivery levels.	_____	_____	_____		

2

Target Market Identification

What You Will Learn

1) The Importance of Target Market Identification
2) How Target Markets Are Identified
3) Key Internal Factors Affecting Target Markets
4) Key External Factors Affecting Target Markets

Operator's Brief

In this chapter, you will learn about the importance of target market identification. A foodservice operation's target market is made up of those who are most likely to first become, and then to remain, customers of the operation. Target market identification is important because, to be efficient, operators want to direct their marketing messages to their most likely customers.

There are a number of ways foodservice operators can segment and then identify their target markets. Four of the most important customer segmentation types are geographic, demographic, psychographic, and behavioral.

Geographic segmentation is based on the location of potential customers. Demographic segmentation is based on shared personal characteristics of customers such as age, gender, and education level. Psychographic segmentation is based on the personality traits, attitudes, values, and shared interests of a customer group. Behavioral segmentation is based on the way members of a target market make their purchasing decisions.

Regardless of the ways operators choose to identify their target markets, they must listen to guests, talk to guests, and maintain accurate sales records to ensure they continue to meet their guests' needs.

(Continued)

As they assess their ability to meet the needs of their target markets, operators must consider both internal and external factors that will directly impact their efforts. Internal factors playing a large role in meeting guests' expectations include an operation's physical facilities and the capabilities of its staff.

External factors influencing an operation's ability to meet its target market's needs include the economic environment, the social environment, and legal requirements. Additional external factors directly affecting an operation's ability to satisfy its target market include vendor and product availability and the competitive environment. Today, technology plays an increasingly important role as foodservice operators meet the needs of their target markets.

CHAPTER OUTLINE

The Importance of Target Market Identification
    Why Target Markets Must Be Identified
    How Target Markets Are Identified
    Identifying the Needs of a Target Market
Internal Factors Affecting Target Markets
    Physical Facilities
    Staff Capabilities
External Factors Affecting Target Markets
    The Economic Environment
    Legal Requirements
    Vendor and Product Availability
    The Competitive Environment
    The Social Environment
    Technology

The Importance of Target Market Identification

Chapter 1 defined marketing as the process foodservice operators use to communicate with their markets; the group of current and potential buyers of the products and services the operators offer for sale.

Some operators may think their market consists of everyone who could possibly buy what they sell. While that may sometimes be true, most foodservice operations know that their **target market** will be most important.

Markets consist of all consumers who *could* buy what an operation sells, but target markets are important because they are the buyers who are *most likely* to become customers. As shown

Key Term

Target market: The group of people with one or more shared characteristics that an operation has identified as most likely customers for its products and services.

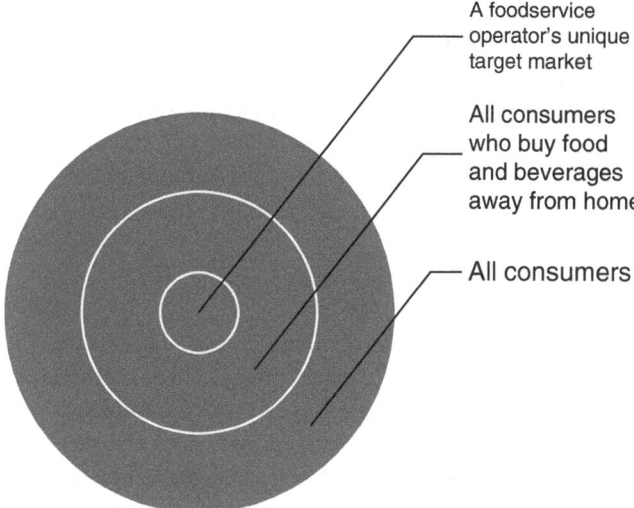

A foodservice operator's unique target market

All consumers who buy food and beverages away from home

All consumers

Figure 2.1   Foodservice Operator's Unique Target Market

in Figure 2.1, a target market is a subsegment of all the consumers who eat while they are away from their homes and will most likely want or need a food operator's products or services.

Why Target Markets Must Be Identified

Target market identification is essential to effective and efficient foodservice marketing. Operators do not want to waste their time and money communicating to consumers who are unlikely to be interested in what the operators sell. Instead, foodservice operators want to carefully tailor their marketing efforts to their target markets.

For example, a very upscale coffee shop located in a downtown's business area is open to serve everyone. This type of coffee shop, however, would likely consider its target market to be those customers willing to pay higher prices for higher-quality coffee products. Not all potential customers are likely willing to do that, at least not every time they purchase coffee.

In this example, the coffee shop's total market may well consist of all persons who are working in or visiting downtown, and who are all coffee drinkers. However, its target market is those individuals who are different from all other customers because they are willing to pay premium prices for premium coffee products.

It is also highly likely that a very large number of guests would enjoy being served high-quality steaks or seafood in an elegant dining room. Foodservice operations serving these high-quality menu items have target markets that include guests who have the desire and willingness to pay, and the ability to pay the higher prices the operation will charge.

To be successful, foodservice operators must carefully identify their target markets and then work diligently to communicate how they will satisfy the wants and needs of that market. When they do so, they will be successful in attracting new customers, retaining their existing customers, and increasing their sales and profits.

How Target Markets Are Identified

Professional foodservice operators use the term "target market" to describe the individuals and businesses that are most likely to buy their products and services. These are the customers with whom operators most want to communicate.

There are several ways to identify a foodservice operation's target market. One common method is to classify target markets by their shared characteristics, and there are four major groupings that are of most importance to foodservice operators:

✓ Geographic
✓ Demographic
✓ Psychographic
✓ Behavioral

Geographic

Target markets can be segmented geographically. Geographic segmentation means identifying a target market based on its location. This is a good strategy when a food operator serves customers in a particular area or when a target market has different preferences based on where they are located. It involves grouping target customers by country, state, region, zip code, city, or even local neighborhood. Segmenting a target market based on its location in an urban, or suburban, area is another example of geographic segmentation.

Geographic segmentation is driven by the principle that people in a specific location may have similar needs, wants, or buying behaviors. For example, a pizza parlor operating near a major university may recognize that its target market includes primarily students who live in, or close to, the university campus. While that pizza parlor may serve nonstudents as well, it is likely that most customers share their geographic location near the university. This common location will likely be more important than gender, income level, or race when the pizza parlor's best potential customers are identified.

Identifying a target market based on geography is especially important when advertising smaller independent foodservice operations because many of these operations may be known only by customers who live or work nearby. Geographic segmentation is important for all operators, however, because it allows them to be more efficient in their communication efforts, and this will reduce advertising and marketing costs.

Find Out More

Professional food-service operators know that identifying a market by geography can do more than simply allow them to better target their advertising efforts. In some cases, a customer's geography will significantly influence the menu items to be offered.

For example, in Hawaii, many McDonald's operators offer Spam as a meat alternative with scrambled eggs, and Thailand McDonald's serves a samurai pork burger. In the United Kingdom, McDonald's serves mozzarella dippers, and in South Africa, McDonald's serves the South African stack. In China, you can buy a McDonald's Cilantro Sundae (vanilla soft serve ice cream covered with lime green sauce and then topped with a hefty pile of dried cilantro leaves).

To see more examples of restaurant operators tailoring their menus to the unique desires of guests found in various geographic locations enter "QSR menu variations by region" in your favorite search engine and view the results.

Demographic

Geographic segmentation separates guests based on *where* they are, while demographic segmentation separates guests based on *who* they are. Some foodservice operators find it best to classify their customers on the basis of demographics.

A target market's demographics consists of a list of statistical data about the characteristics of a specific group of people. In simple terms, a demographic is one characteristic of a larger population.

Examples of demographic characteristics of a potential target market include:

✓ Age	✓ Occupation	✓ Nationality
✓ Sex	✓ Education level	✓ Religion
✓ Marital status	✓ Income	
✓ Family size	✓ Race	

In some cases, a single demographic factor can completely include, or exclude, a potential group of customers. This is true, for example, for foodservice operators who sell alcoholic beverages in addition to food. In this example, only those guests who are of legal drinking age are potential customers for the operation's alcoholic beverage items, while a much wider range of customers, with a different "age" demographic characteristic, can be buyers of the operation's other menu items.

Similarly, some foodservice operations offer children's menus only to customers younger than a specifically identified age. Both these examples show how a demographic characteristic (age) can help determine an operation's target market.

Other examples of directly addressing customers' demographic characteristics include foodservice operations that promote their kosher menu items (items prepared for those who follow Jewish dietary practices) and halal menu items (items prepared for those who follow Muslim dietary practices). Given the increasing multiculturalism of Americans, it is becoming more common for foodservice operators to identify these two major religious groups as target markets based on the demographic characteristic of religion.

Psychographic

Psychographic segmentation is similar to demographic segmentation, but it addresses characteristics of a target market that are mental or emotional. Psychographic characteristics include customers' personality traits, attitudes, values, interests, and beliefs.

These attributes are not easily seen; however, they can give valuable insight into a target market's motives, needs, and preferences. While demographics provide factual characteristics of "who" their customers are, psychographics give foodservice operators insight into "why" their customers make their buying decisions.

One good example in the foodservice industry of appealing to a psychographic characteristic are those operations that advertised themselves as being "eco-friendly." In this example, the operations seek to communicate their pro-environment business philosophy to a select group of people who feel it is important to practice sustainable and **green practices** when operating a business. These individuals could exist in several different geographic and demographic classifications.

Those foodservice operations providing "Early Bird" specials are another example of appealing to a psychographic characteristic. In this case, the customer characteristic of preferring to dine early in the evening is targeted by operators. Food and beverage operations offering "Happy Hour" drink specials provide another example of utilizing a psychographic characteristic (i.e., customers seeking lower drink prices if they buy at a specific time of day) to target potential customers.

> **Key Term**
>
> **Green practices:** Those activities that lead to more environmentally friendly and ecologically responsible business decisions. Also known as "eco-friendly" or "earth-friendly" practices.

Behavioral

Foodservice customers can also be targeted on a behavioral basis. For example, operators can categorize their potential customers based on whether those customers are frequent buyers, infrequent buyers, or nonbuyers of their products and services.

Some of foodservice operations customers will be very loyal, while others can be easily persuaded to visit competitive operations. Marketing efforts directed at frequent buyers are very common in the foodservice industry.

The rewards can be as simple as a free cup of coffee or as valuable as a completely free dinner. Customer loyalty programs based on the behavior of customers are a very important component of foodservice marketing today, and these loyalty-type programs will be discussed in greater detail in Chapter 8.

Technology at Work

Many foodservice operations use guest loyalty programs as part of their marketing strategy. Essentially, these programs offer those customers who already enjoy a foodservice operation's products and services some additional incentives to return often.

Typically, a customer loyalty program offers guests a discount, a free menu item, or specific merchandise as a reward for frequent buying of qualifying purchases.

In the early days of foodservice rewards programs, customers were typically given a punch card to record the number of times they had visited or the total value of purchases over time. Today specially designed customer loyalty program software does away with the need for such cards. Since they are easy to initiate and maintain, nearly every foodservice operator today should utilize a customer loyalty program.

To review specific guest loyalty program software and apps appropriate for use in a foodservice operation, enter "foodservice customer loyalty program software" in your favorite search engine and view the results.

Today, very inexpensive computer apps allow an operation to enter their customers' phone numbers into their **point-of-sale (POS) system** whenever a purchase is made, and specific rewards can be given to the customer.

A second example of buyer behavior impacting marketing efforts relates to potential customers who seek information online versus those

Key Term

Point-of-sale (POS) system: An electronic system that records foodservice customer purchases and payments, as well as other operational data.

who are not frequent seekers of online information. While operators must be careful not to overgeneralize. However, many younger consumers are comfortable seeking information about a foodservice operation online or with various smart device apps. Alternatively, many older consumers may be less comfortable doing so. The COVID-19 pandemic of 2020–2022 created numerous changes in the way foodservice operators marketed their products and services. As foodservice operations were locked down, and customers started to order and buy their food online, those operations that did not have a significant online presence suffered greatly.

Alternatively, those foodservice operators who were able to communicate with their tech-savvy targeted guests with online placement of their menus and photos of their food offerings, were more successful in reducing some of the negative impact of the pandemic.

Regardless of the segmentation approach used, the identification of target markets is critical to all foodservice operators. Some foodservice operations may have multiple target markets. That is, they may have a primary target market which is their main focus, and a secondary target market which may not be as large, but which may have great growth potential.

In all cases, foodservice operators who understand their target markets truly are, or could be, will be in a better position to devise effective **marketing messages** geared directly toward those target markets.

Key Term

Marketing message: How an operation communicates to its customers and highlights the value of its products and services.

Find Out More

Professional marketers know that one of the most significant differences in the buying behavior of all consumers, including foodservice consumers, relates to the year of their birth.

While the exact birth dates identified can vary somewhat, most professional marketers would agree that the following generations share specific characteristics as consumers:

✓ Baby Boomers (born 1946–1964)
✓ Generation X (born 1965–1980)
✓ Millennials (born 1981–1995)
✓ Generation Z (born 1996–2010)
✓ Generation Alpha (born 2011–2025)

To cite just a few examples of significant generation-related differences of importance to foodservice operators, a study by Kognitiv Internal Benchmarking

(2019)* found that, because many have families to provide for, Generation Xers are keen on finding good deals that will make their life easier, and they are, in fact, the generation most likely to redeem gift cards in a loyalty program.

The same study found that Millennials are heavy social media users (43% use social networks to fill their spare time). This means they are highly susceptible to social proof from friends and trusted individuals when making purchase selections.

To review additional and current behavioral differences reported among the generations enter "foodservice market segmentation by generation" in your favorite search engine and view the results.

Source: https://kognitiv.com/blog/generational-loyalty-from-boomers-to-gen-z/ retrieved April 15, 2022.

Identifying the Needs of a Target Market

After a foodservice operation has identified its target market, it must consider the specific needs, wants, and desires of that market. A customer's need can be defined simply as something they lack, or the difference between where the customer is and where they would like to be.

For example, a customer may be hungry (what they currently are) and they want to be well-fed (what they would like to be). This need for food would cause them to seek out a foodservice operation where they hope their need would be satisfied. If that customer's meal purchase met their expectations, then they would likely want to return to the operation again in the future. If, however, they were not satisfied (the meal did not meet their expectations), they would not likely return in the future. It is important to remember that successful marketing involves first identifying and then meeting, or exceeding, the expectations of customers.

Hunger is one need that could cause a customer to seek out of foodservice operation, but it is not the only need. If foodservice operations existed only to meet hunger needs, they might all look virtually the same and offer the same menu items. However, foodservice guests have many more needs than simply satisfying their hunger.

To illustrate, assume that a foodservice operation has determined that senior citizens are a major target market. While these customers certainly desire well-prepared menu items, they are also likely to desire a welcoming environment, more lighting, less noise, easy-to-read (large-type) menus, comfortable seating, and high levels of personal attention. Each of these needs is in addition to their need to satisfy their hunger.

In a similar manner, cost-conscious customers utilizing the drive-thru lane of a quick-service restaurant certainly desire well-prepared menu items. However, they are likely to also be very interested in their order's speed of delivery, accuracy in filling their orders, and appropriate packaging of all their selected to-go menu items. Each of these desires will be in addition to their need to satisfy their hunger at a relatively low price.

When target markets have been identified, the needs and desires of these specific markets can also be identified. Foodservice operators have a number of ways in which they can identify their target markets' needs. Three of the most important of these are to:

1) Listen to customers
2) Ask questions
3) Keep good sales records

Listen to Customers

Most foodservice operators understand the importance of interacting directly with their customers. Those interactions are most helpful, however, when operators carefully listen to what their customers are saying. In most cases, customers will not be hesitant to tell what they like and dislike about an operation's prices, menu items, and service levels.

It is important for foodservice operators to listen to their customers for good and bad comments. If there are menu items or service procedures causing dissatisfaction among a target market, it is important for operators to become aware of that so they can make improvements.

One common way to find out what customers think about an operation is to circulate in dining areas and directly engage guests in conversation. Today, increasingly, foodservice operators will find out what their customers are saying by reading the reviews these customers leave on online review sites. This nontraditional way of listening is especially important because customers sometimes are more open and honest when they are posting a review online than when they are talking to a manager or server in person.

Ask Questions

Even the best of foodservice operators are not mind readers. So, one good way to find out about the needs and wants of target customers is to simply ask them. Asking guests about their needs and wants not only provides valuable information but it also communicates to the target market that management genuinely cares about them and how they can best serve them. Increased customer loyalty is often a positive by-product of conversations that demonstrate an owner or manager is truly interested in their customers' opinions.

In addition to holding conversations with guests, an extremely popular way to gather information about their wants and needs is through surveys. These surveys need not be complex, and guests may be asked to fill them out when they are on-premise or online.

Typical questions that can provide valuable information include:

✓ How did you find out about our operation?
✓ How often do you come to our operation?
✓ How did you place your order with us?
✓ How would you rate the speed of our service?
✓ What items would you like to see us add to our menu?
✓ How would you rate the quality of our food?
✓ How would you rate the quality of our service?
✓ Do you feel that you get good value for the money you spend on our operation?
✓ How likely would you be to recommend our operation to your friends and family?
✓ What changes would you suggest to make our products and services even better for you?

The best foodservice operators not only survey their guests regularly, but they carefully record and analyze survey information. They also pay particular attention to those areas in which many members of their target markets indicate the same response, whether it is positive or negative.

Keep Good Sales Records

It is important to keep good records about what the foodservice operation's customers actually buy. This important topic will be addressed in detail in Chapter 6, but from the perspective of satisfying the needs of an operation's target market, sales records indicating what menu items are selling, and when they are sold, can provide important insight into the purchase behavior of the target market.

When the sales of an individual menu item are increasing, it can indicate that the market is increasingly satisfied with that item. Alternatively, if a menu item's sales are low, or decreasing, it can indicate that this item is not meeting the needs of the operation's target market.

In some areas of foodservice, such as in hospitals, retirement centers, and smaller college and university residence halls, the number of menu items offered to guests on any specific day may be quite limited. Then it may be less insightful to monitor the number of times a specific menu item is served to guests. Instead, it is especially important that the operators of these facilities talk and listen to guests as they continually monitor menu item preferences.

In addition to keeping good records about what menu items are selling (and when they are selling), it is increasingly important to keep good records about the manner in which the items are being sold. For example, in the not-too-distant past, most foodservice operations sold the majority of their menu items to on-premise diners. More recently, many foodservice operations sell increasing numbers of their items via drive-through, pick-up, or third-party delivery. When these options are offered, it is the best practice for foodservice operators to monitor their menu item sales based on the ways the items are actually delivered or served to their guests.

Technology at Work

A large number of companies sell foodservice industry-specific POS systems to record individual menu item sales information and do much more. Some of the most popular of these companies currently include:

Revel Systems
Toast
Squirrel Systems
Touch Bistro
Oracle
Lightspeed
Clover (for food trucks)

The proper analysis of individual menu item sales is a powerful management tool, and it is particularly helpful for operators in analyzing their product mix and learning what is selling well and what is not.

To review the type of menu item sales tracking reports provided by each company, enter the company's name followed by "POS sales tracking" in your favorite search engine and view the results.

What Would You Do? 2.1

"That's the third customer today that left without buying anything," said Carla. Carla was talking to Lamont, the owner of The Tasty Bakery; a small bakeshop that specialized in premium-quality cakes, cookies, muffins, and pastries.

"Did they think our prices were too high?" asked Lamont.

"No, they didn't seem to have any problem with the prices," replied Carla, "They wanted a dozen gluten-free chocolate chip cookies for an office party this afternoon. I told her the only gluten-free item we sold was our blueberry muffins. The other two customers who left without buying anything today were looking for gluten-free cookies as well."

Assume you were Lamont. How important do you think it would be before you to carefully monitor your customers' requests for items they want to buy from your shop but cannot? How would you learn about their interests?

Internal Factors Affecting Target Markets

After identifying their target markets and assessing their markets' needs, operators will want to offer the products and services these markets seek. However, a number of internal and external factors will directly affect their ability to do so.

The most important of the internal factors affecting an operator's ability to meet its guests' needs and that must be carefully assessed are the operation's physical facility and its staff capabilities.

Physical Facilities

An operation's physical facility will directly impact its ability to target specific markets and meet its target markets' needs. Limitations, or constraints, on an operation's ability to do this well can be related to both front-of-house and back-of-house issues.

In the front of the house, it is easy to see that foodservice operation may be limited in the total number of guests it can serve based on the number of seats available in the facility. For busy operations, this front-of-house capacity-related issue is best addressed by efficient table layout, the use of an effective **table management system**, and proper front-of-house staffing.

Key Term

Table management system:
A software program, interfaced with a foodservice operation's POS system, that efficiently manages the operation's guest reservation and seating processes.

Technology at Work

Table management systems are designed to help operators make efficient use of their available seating during busy service periods.

The best of these systems allows operators to:

✓ Load in a diagram of their specific dining room layouts
✓ Monitor guest reservations made and their arrival times
✓ See the number of guests seated at any point in time
✓ View all open tables and seats
✓ Estimate dining times and forecast seat availability
✓ Track payment status by table

To see the various features offered in a variety of popular programs, enter "table management software" in your favorite search engine and view the results.

Those operations that offer drive-through and pick-up services can also face significant front-of-house capacity issues. While many diners enjoy a leisurely meal, most QSR customers using drive-through lanes want their menu selections delivered accurately and, most importantly, they want their orders delivered quickly. During busy service times, these goals can be a significant challenge.

Segregating a separate lane for delivery drivers and customers that order via mobile devices can make lanes flow more efficiently. When customers order via a mobile app, they give the kitchen staff advance time to prepare the orders, and by paying in advance, they eliminate the need for staff to process payments. These mobile advantages can yield significant time savings and speed up the delivery of customer orders. In addition, touchless payment options such as **mobile wallets**, apps that hold preloaded credit card information, and one-touch credit card readers, can speed up service through conventional drive-through lanes.

In some cases, a foodservice operation can increase the number of drive-throughs it operates but, in others, space limitations may mean that it cannot do so. Then operators can consider dispatching order-takers into drive-through lines at peak times of the day. With this approach, staff members can walk out to the drive-thru lane and, using a tablet integrated with their in-house POS system, they can take customer orders to expedite order preparation, payment, and delivery.

Back-of-house limitations related to target markets typically involve production equipment capacities. These should be rare in a well-designed kitchen, however, in some cases, such as in food trucks, back-of-house (kitchen) capacities may be relatively fixed based on the size of the truck and the number of production staff it can accommodate.

In all cases, foodservice operators must assess the physical facilities they have in place, and any limitations imposed on those facilities, as they seek to satisfy the needs of their selected target markets.

Key Term

Mobile wallet: A virtual wallet that stores payment card information on a mobile device. Mobile wallets provide a convenient way for a user to make in-store payments and can be used at merchants listed with the mobile wallet service provider. Also known as a digital wallet.

Find Out More

As drive-throughs continue their increase in popularity among foodservice customers, operators are increasingly concerned about monitoring the time it takes to process a drive-through order because most drive-through customers are in a hurry.

A study conducted by SeeLevel HX; a customer experience measurement company recently found wait times for receiving a drive-thru order increased by more than 25 seconds from 2020 to 2021.*

The study classified the total time customers wait for an order from the moment they enter the drive-through to the moment they got their order. The study found that in 2021, drive-through customers waited an average of 6 minutes, 22 seconds. In 2020, it was 5 minutes, 57 seconds. The research involved nearly 1,500 drive-thru visits between July and August 2021 to 10 major fast-food brands including McDonald's, Burger King, Taco Bell, Chick-fil-A, and Dunkin (formerly Dunkin Donuts).

Wise foodservice operators will no doubt continue to carefully monitor the average amount of time it takes to serve their customers through their drive-throughs. To find up-to-date information about drive-through service times, enter "average foodservice drive-through speeds," in your favorite search engine and view the results.

* https://www.prnewswire.com/news-releases/seelevel-hx-21st-annual-drive-thru-study-uncovers-delays-and-inaccuracy-as-qsrs-struggle-with-labor-shortage-301383881.html retrieved May 1, 2022

Staff Capabilities

Regardless of the type of operation, meeting the needs of a target market requires consistent excellence in guest service. Service excellence requires the best efforts of every foodservice employee, from manager to order taker to dishwasher. Unless every person does his job properly, meeting target market guest needs will be challenging.

Finding and keeping qualified staff members is more difficult in some labor markets than in others. Regardless of the challenges involved in securing staff, foodservice operators can help minimize the staffing limitations they may face by following employee-focused best practices. These include:

1) Hiring for attitude rather than experience

Experience is important, but an inexperienced worker with a positive attitude can often make for a better staff member than one who has been on the job longer but lacks the passion to serve guests. Attitude cannot be taught, but job skills can. In the best possible case, a potential employee will have both relevant experience and a positive attitude, but staff members with a positive guest service attitude should not be dismissed as job candidates simply due to limited experience.

2) Leading by example

The guest service attitude of a foodservice operation's managers set the tone within the operation. If a foodservice operation's employees are to treat guests with dignity and respect, they must be treated the same way by their operation's managers. The ways that foodservice operators treat their staff will also reflect how staff members treat each other. Managers consistently modeling and displaying a positive attitude at work will make that the norm for staff, rather than the exception.

3) Providing fair wages and benefits

In many cases, the competitors for quality foodservice staff are not just other foodservice operations, but rather all service organizations seeking qualified workers. While it is true that some jobs in foodservice are relatively low-paying, that does not mean that operators cannot reward staff in other creative ways. This can include providing employee meals, flexible scheduling, and special incentives for longevity.

4) Listening to staff

A foodservice operation's staff members interact with customers directly. As a result, they are an excellent source of knowledge about what a target market is thinking and saying about an operation's products and services. When foodservice operators seek input from their full- and part-time staff members who are actually serving guests, the operation as a whole gains as it is better able to meet its target market's needs. When staff members feel their opinions are valued, it also makes them feel more valued as team members and their work performance improves.

5) Making staff training a priority

To provide the best quality service to guests, employees must be well-trained. Regardless of the style of operation they are visiting, most guests go out to eat or place home delivery orders because they want a pleasant dining experience. As a result, excellence in service is a core factor of guest satisfaction.

Well-trained service staff speak courteously to guests and go out of their way to provide special service. To do that they must first be trained. A well-trained workforce saves foodservice operations time and money, and it increases their profitability because it encourages first time customers to become return customers.

6) Cross-training capable staff

Cross-training those staff who are talented enough to benefit from it will improve the entire foodservice operation. Effective foodservice operators cross-train their staff so that one worker can fill multiple roles.

Key Term

Cross-training: The action or practice of training workers in more than one role or skill.

When significant numbers of staff are cross-trained in various positions, a foodservice operator is better prepared to deal with employee vacation schedules, call-ins, and no-shows.

7) Scheduling professionally and properly

In too many foodservice operations, improper employee scheduling is a consistent source of worker dissatisfaction. When too few workers are scheduled to meet the demands of guests, guest service suffers and employee frustration increases. When too many workers are scheduled, staffing costs are unnecessarily increased.

The best foodservice operators involve their employees in the scheduling process. Giving hourly workers a voice in the scheduling process shows operators care enough to ask staff about their work time preferences. When this information is known ahead of time, it allows managers in a foodservice operation to build schedules with a high degree of staff acceptance.

To help ensure work schedules are properly developed, it is also a good idea to have another manager or senior staff member check them before they are distributed to staff.

8) Communicating work schedules well in advance

Studies of worker satisfaction in the foodservice industry consistently show that many workers find out their work schedules a week or less in advance. Not knowing when they will be scheduled to work can make it very difficult and, especially for part-time workers, to plan the rest of their lives and their other work schedules.

In most cases a foodservice operator will have a very good idea of customer demand much further than one week in advance, and schedules should be made and distributed as early as possible. Last-minute modifications to schedules are also easier to make when workers have had enough lead time to plan around their other activities.

9) Allowing for maximum schedule flexibility

Just as the foodservice industry can be unpredictable regarding guest demand, foodservice staff members' lives can be unpredictable as well. The best foodservice operators recognize this fact and allow their staff members as much flexibility as possible when scheduling. This can include, as needed, trading shifts with other workers who are qualified to accept them.

It is important to note that restaurant workers, and especially those paid hourly, are increasingly, leaving the foodservice industry due to what is now popularly referred to as the **gig economy**. In most

Key Term

Gig economy: A labor market characterized by the prevalence of short-term contracts or freelance work as opposed to more permanent jobs.

cases, the gig economy is attracting former foodservice workers due to the flexibility of work arrangements it provides.

While not all foodservice operations can offer maximum flexibility when scheduling their staff, all foodservice operators should carefully consider how flexible scheduling can help retain their staff.

10) Empowering staff to make it right

Regardless of their best efforts, experienced foodservice operators know that guest complaints will occur. That is inevitable. How those complaints are addressed, however, is well within the control of every foodservice operator to control.

One of the best ways to help ensure target market satisfaction is by the **empowerment** of an operation's service staff.

When staff members are allowed to use their good judgment when addressing service shortcomings, guest complaints can be dealt with more quickly and face to face. Granting qualified staff members permission to handle problems as they occur dem-

Key Term

Empowerment (staff): An operating philosophy that emphasizes the importance of allowing staff to make independent guest service-related decisions and to act on them.

onstrates an operator's faith in the staff members and confidence that the staff will do the right thing for guests and the operation.

External Factors Affecting Target Markets

Although foodservice operators have the ability to select their target markets and ensure their products and services are designed to meet their target markets' needs, a number of important external factors will influence their efforts. These include:

✓ The Economic Environment
✓ Legal Requirements
✓ Vendor and Product Availability
✓ The Competitive Environment
✓ The Social Environment
✓ Technology

The Economic Environment

A goal of marketing is to help operators identify and satisfy the needs of their target markets. The ability of a target market to make purchases is directly affected by the economic health of the city, state, and country in which it is located. Unemployment rates, inflation, personal income growth, and confidence in the economy are all factors that directly impact the **disposable income**, and thus the **discretionary income** of consumers.

Foodservice operators should regularly monitor their local economies and assess any impact it may have on their target markets because the health of the economy directly impacts the amount these markets have available to spend. During good economic times, consumers are willing to spend more, while in times of economic difficulties, the number of potential customers and the amount of discretionary income available to them may decline.

Key Term

Disposable income: The amount of income left over after all taxes are paid. Also known as net personal income.

Key Term

Discretionary income: The amount of disposable income remaining after all basic necessities such as housing, food, and clothing are paid.

Legal Requirements

At all levels of government, including local, state, and national, laws and regulations affect how a foodservice operation can serve its target markets. Restrictions on the sale of alcohol, prohibitions against indoor smoking, and **accuracy in menu laws** are just a few examples of legislation that operators must comply with regardless of the desires of some of their customers. Also, at the Federal level, laws are in place that make it illegal to withhold service from anyone based on their race, religion, or national origin.

Foodservice operators must ensure that they are in compliance with all laws affecting the operation of their businesses. They should also make their voices known at the appropriate governmental levels when they feel newly proposed regulations will help, or harm, their businesses.

Key Term

Accuracy in menu laws: Legislation that requires foodservice operations to represent the quality, quantity, nutritional value, and price of the items they sell truthfully and accurately.

Also known as "Truth-in-menu laws" and "Truth-in-dining laws."

Find Out More

Foodservice operators are granted wide latitude in how they describe and market their products and services. However, misrepresenting what is offered for the purpose of deceiving customers is illegal and unethical. Inaccurate or incomplete menu item descriptions can also lead to injury and illness affecting those guests who have food allergies.

(Continued)

> Legislation in this area is extensive and wise foodservice operators can stay abreast of new laws and regulations by membership in professional associations and by their own research.
>
> To learn more about what foodservice operators can, and cannot, say about their products and the prices charged for them, enter "accuracy in menu legislation," in your favorite search engine and view the results.

Vendor and Product Availability

An operator's ability to meet its target market's product needs is essential for guest satisfaction.

In many cases, members of a target market choose the foodservice operations they will visit based on the menu items offered for sale. Therefore, it is essential that foodservice operators are able to consistently receive the quality and quantity of raw ingredients needed to produce their featured menu items.

In larger communities, operators may have a wide choice of vendors, thus helping to ensure product availability from a variety of sources. In smaller communities or more remote locations, an operator's vendor choices may be quite limited. In all cases, ensuring product availability is an external factor that will play an important role in an operation's ability to consistently meet its guests' needs.

The Competitive Environment

Nearly every foodservice operation will face competition for its target market. Experienced foodservice operators know that there is typically a relatively fixed amount of discretionary income available to spend in a market area, and there may be many foodservice operations vying for those discretionary dollars. To stay in business, a foodservice operation must meet the needs of its target market as well or better than its competitors.

There are a number of ways in which foodservice operators can assess their competitive environment. One good way to assess competitors is to group them based on one of three operating characteristics. These will be those operations that offer guests:

✓ Similar products
✓ Similar service style
✓ Similar prices

Similar Products

It is easy to see that those foodservice operations offering similar products will be an operation's direct competitors. For example, in the QSR segment, McDonald's,

Wendy's, Burger King, Sonic Drive-In, Jack in the Box, Whataburger, and more compete directly for customers seeking to buy modestly priced hamburgers, French fries and related menu items. Similarly, Dominos, Pizza Hut, Little Caesars, Papa John's, and other pizza **chains** all compete directly for those customers seeking to buy modestly priced pizzas. Even very upscale restaurants serving steak and seafood in an elegant setting will likely encounter other upscale operations in their market area serving steak and seafood in similar settings.

Regardless of the segment in which they operate, in most cases, a foodservice operation will face direct competitors offering very similar products. Understanding what these direct competitors offer their guests, how their products are delivered, and their pricing structure is imperative to understanding the competitive environment.

Key Term

Chain (restaurant): A group of restaurants with many different locations that share the same name and concept. Chain restaurants may be owned and operated by the chain, and/or they may be individually owned through franchising. Chain restaurants may be classified as QSRs, fast casual, casual or up-scale operations.

Similar Service Style

McDonald's competes directly with Wendy's, but it also competes with Chick-fil-A, Taco Bell, and Pizza Hut. Each of these chains offers modestly priced menu items for purchase to-go or for dine-in. When targeting those diners seeking low-cost, quick-to-purchase and pick-up menu items, all chains and independent operations offering these same types of items in the same service style would be considered competition.

In the same way, those foodservice operations offering their guests mid-priced on-premise dining options, whether selling steaks, seafood, Mexican, Italian, or other menu options, should be considered direct competitors if they offer the same or similar on-premise service style.

Similar Prices

In many cases, a decision about whether to frequent a specific foodservice operation is based on that operation's pricing structure. In most markets, the average foodservice consumer can choose from low-cost, mid-price, or very upscale dining options. Special occasions and celebrations may cause a foodservice consumer to be willing to spend more than they would normally spend. Alternatively, having a limited amount of time to eat may indicate the same consumer's desire for a lower-cost option that can be delivered quickly.

Of course, those consumers with higher discretionary income will most often view this decision differently from those with lower levels of income. But, in an overwhelming number of cases, customers make decisions about the amount they are willing to spend *before* they assess their dining-out options. As a result, a food-service operator should consider all of those operations offering its target market similarly priced options when it assesses its direct competitors.

Find Out More

Popeyes Louisiana Kitchen (Popeyes) is an American QSR chain featuring fried chicken. As of 2020, Popeyes operated 3,450 restaurants located in more than 46 states and the District of Columbia, Puerto Rico, and 30 countries worldwide. Popeyes was founded in New Orleans, Louisiana in 1972. It is currently a subsidiary of Toronto-based Restaurant Brands International, who are also the owners of the Burger King, Tim Hortons, and Firehouse Subs brands.

When Popeyes first introduced its chicken sandwich in August 2019, customer demand for it was so overwhelming that the sandwich was sold out chain-wide after only two weeks. The sandwich was reintroduced in the Fall of 2019 after the chain had a chance to secure enough products, and add enough new employees, to meet the demand for this game-chaining new menu item.

Seeing the response of customers, nearly every other QSR chain scrambled to develop and introduce its own competitive version of Popeye's innovative chicken sandwich. Some were moderately successful, and others much less so, but nearly all major chains felt they needed a rapid product response to Popeye's incredibly popular new menu offering.

To learn more about this excellent example of the impact of competitors on a foodservice operation's own menu decisions, enter "response to Popeye's chicken sandwich rollout," in your favorite search engine and view the results.

The foodservice industry is highly competitive. New operations open, and others close, on a regular basis. For that reason, it is important that food-service operators continually monitor, and stay up-to-date about who their competitors are, and what those competitors are offering to their potential target markets.

The Social Environment

Societal views on a number of issues directly affecting foodservice operations continue to evolve and should be monitored. These changing views impact the items a foodservice operator chooses to sell, how they are sold, and the marketing messages operators must use to sell them.

For example, consumers, including foodservice customers, increasingly feel it is important that the businesses they frequent are instituting green practices. As a result, how operators source their foods, prepare them, and package them for sale has undergone significant change.

To cite just one example of green practices directly impacting foodservice operations, in June 2022, Starbucks reintroduced personal reusable cups across all of its U.S. stores. The practice of allowing guests to bring in their own cups for refilling was one part of Starbucks' ongoing commitment to reduce single-use cup waste, as well as to respond to its target market's desire to embrace earth-friendly packaging.

Other examples of green practices that may become part of a foodservice operation's marketing message include energy conservation efforts, adopting **farm-to-fork** product sourcing methods, and instituting recycling programs.

The increasing popularity of plant-based meat products is yet another example of why operators must continually monitor changing societal views that may directly impact their product offerings and target market messages.

Key Term

Farm-to-fork: A procurement process that stresses the importance of cooking with the freshest, locally, and sustainably grown seasonal ingredients. Also known as "farm-to-table."

Technology

Perhaps no other external factor has impacted foodservice operators more than changing technology. There is no doubt that technological advancements will continue in a variety of important areas of foodservice. Some recent examples of technology change that have spread industry-wide and have a direct impact on marketing include:

✓ Mobile ordering
✓ Self-service kiosks
✓ Cloud-based POS systems
✓ New payment options
✓ Online reviews

Mobile Ordering

Due in part to the impact of the COVID-19 pandemic, and in part because of changing consumer preferences, mobile ordering has become the norm for many foodservice customers. Mobile ordering reduces wait times for guests in a hurry and for those who simply like its convenience.

Self-Service Kiosks

Already in use by many QSR chains, self-service kiosks are designed to speed order entry and improve order accuracy. These units also help address labor shortages, a common problem among some foodservice operators.

Cloud-based POS Systems

Cloud-based POS systems are becoming increasingly popular because they are easy to install and required software updates can be done remotely. Also, the data management capabilities of most cloud-based POS systems are superior to those of hard-wired systems, giving operators who choose them more information for use in menu planning, staffing, and financial reporting.

New Payment Options

Foodservice guests paying for their meals via phone app, online portal, or **QR code** has increasingly become commonplace. These options are growing in popularity because they simplify and speed up the payment process.

Online Reviews

In the past, the impact of a product or service error made in a foodservice operation was limited to the guest involved, and perhaps those that the guest would personally tell about the experience. Today, online restaurant review sites allow all guests to comment on their experiences to virtually millions of potential readers. These sites are very popular with many guests and, as a result, foodservice operators must ensure their product and service delivery is excellent. They must also continually monitor these sites to see how they are viewed by their guests, and they must respond professionally when they need to do so (see Chapter 11).

Key Term

QR code: A QR (quick response) code is a machine-readable bar code that, when read by the proper smart device, allows foodservice guests to view an online menu, be redirected to an online ordering website or app, or that allows them to order and/or pay for their meal without having to interact directly with a foodservice operation's staff.

In all of these technology-related examples, subsections of an operation's target market may be directly impacted. As a result, operators must closely monitor all of the technological advancements their target markets have adopted or will soon adopt.

It is essential that all foodservice operators identify their target markets. A suitable target market is one that:

✓ can be well-defined with readily identifiable geographic, demographic, psychographic, or behavioral characteristics.
✓ is large enough to meet the operation's revenue and profit goals.
✓ is stable or growing in size.
✓ has a need that can be filled by an operation's products and services.
✓ is accessible and is able to be reached with the operation's marketing budget and marketing message.

After a foodservice operation has identified its target market(s), it can begin the important task of creating its marketing message, and that process is the topic of this book's next chapter.

What Would You Do? 2.2

"This seems pretty complicated," said Sherrie, as she peered closely at her cell phone.

"What could be so complicated?" asked her roommate Kim. "It's just ordering a pizza."

"Well, I know, but they have thin crust, thick crust, and deep-dish choices. I want to order us a medium Supreme, thin crust, but with extra cheese and no black olives. I don't see how I tell them all that," replied Sherrie.

"Maybe you should just call them then," said Kim, "that might be easier."

Assume you were the operator of this pizza shop. What steps could you take to ensure the introduction of an online ordering app for your target market creates a positive guest experience rather than a frustrating one? How important would it be for you to do so?

Key Terms

Target market	Table management	Disposable income
Green practices	system	Discretionary income
Point-of-sale	Mobile wallet	Accuracy in menu laws
(POS) system	Cross-training	Chain (restaurant)
Marketing	Gig economy	Farm-to-fork
message	Empowerment (staff)	QR code

Operator's 10-Point Tactics for Success Checklist

Evaluate your need for, and the current status of, each of the following operational tactics. For those tactics you think are important, but not yet in place, develop an action plan for its implementation including who will be responsible for the tactic's completion and the target date by which it should be completed.

Tactic	Don't Agree (Not Done)	Agree (Done)	Agree (Not Done)	Who Is Responsible?	Target Completion Date
				If Not Done	
1) Operator has carefully considered the impact of geography on the targeted market.	___	___	___		
2) Operator has carefully considered the impact of guest demographics on the targeted market.	___	___	___		
3) Operator has carefully considered the impact of psychographic factors on the targeted market.	___	___	___		
4) Operator has carefully considered the impact of behavioral factors on the target market.	___	___	___		
5) Operator has assessed and addressed all front-of-house and back-of-house facility limitations that may impact the target market.	___	___	___		
6) Operator has assessed and properly addressed all staffing limitations that may impact the target market.	___	___	___		
7) Operator has carefully considered the manner in which any economic, legal, and vendor-related issues may impact the target market.	___	___	___		

Tactic	Don't Agree (Not Done)	Agree (Done)	Agree (Not Done)	If Not Done	
				Who Is Responsible?	Target Completion Date
8) Operator has carefully assessed each competitor vying for the same target market.	——	——	——		
9) Operator has carefully considered the manner in which the social environment may impact the target market.	——	——	——		
10) Operator has assessed the manner in which technology may impact the target market.	——	——	——		

3

Creating the Marketing Message

Operator's Brief

In this chapter, you will learn about the importance of a foodservice operation's marketing message. The marketing message is a key to operational success because it communicates to guests the reasons why they should become and remain good customers. With an effective marketing message, guests clearly understand what they will receive and how they will benefit when they choose a specific foodservice operation.

The marketing message is one major component of branding. Branding is the way in which a foodservice operation can tell its unique story. Effective branding allows a foodservice operation to tell its potential guests who it is, what it stands for, and what it promises to deliver. Through its use of brand identifiers including brand marks and trademarks, an operation must present a consistent and easy-to-understand message. In most cases, operators choose to emphasize either their product offerings or their service style as they develop their marketing message.

To emphasize product offerings as brand identifiers, operators can choose several different approaches. These can include product quality, serving size, price, convenience, and location. Any one, or a combination of these focus areas, can help a foodservice operation differentiate itself from its competitors.

When deciding to emphasize service level as a brand identifier operators can also choose from several different approaches depending on their classification. Thus, non-commercial foodservice operations, Ghost operations, and Quick Service Restaurants (QSRs) might decide to emphasize their service areas. Similarly, fast casual restaurants, casual service restaurants, and fine dining restaurants may also select service level as a major brand identifier.

In all cases, brand identifiers are utilized to make an operation stand out from its competitors by clearly stating in their marketing messages who they are and what guests can expect when they visit the operation.

CHAPTER OUTLINE

The Marketing Message and Importance of Branding
    The Marketing Message
    Branding Defined
    Developing a Unique Brand
    Product-Focused Branding
    Service-Focused Branding
Factors Affecting Product-Related Messages
    Product Quality
    Preparation Method
    Serving Size
    Price
    Convenience
    Location
Service Offering Messages by Type of Operation
    Non-commercial Foodservice Operations
    Quick Service Restaurants (QSRs)
    Fast Casual Restaurants
    Casual Service Restaurants
    Fine Dining/Upscale Restaurants
    Ghost Operations

The Marketing Message and Importance of Branding

Chapter 2 defined the marketing message as the way an operation communicates to its customers and highlights the value proposition of its products and services. The specific marketing message used to communicate with a target market will vary from one operation to the next, but to be effective the message must be clear and consistently delivered.

The Marketing Message

The owners and managers of every foodservice operation would likely believe that they provide good food and good service. But that alone cannot be their marketing message because it does not tell their target markets exactly why their specific operations should be chosen among all other possible choices.

An effective marketing message represents how an operating **brand** communicates to its customers and highlights the value of its products and services, and it also conveys the feelings and emotions associated with the brand.

Rather than highlight any one specific product or service, brand marketing promotes the entirety of the brand. Then, an operation's specific products and services offered provide the proof that will support the brand's promises. The marketing message is just one important distinguishing feature of branding.

Key Term

Brand: The specific ways in which a foodservice operation differentiates itself from its competitors.

Branding Defined

A brand can be thought of as the personality of a foodservice operation as communicated through its name, logo, and any of a large number of other **brand identifiers**. A good brand identity attracts new guests to a foodservice operation while continuing to make existing guests feel good about choosing the operation again.

The proper marketing of an operation's brand (branding) is becoming more important than ever in the foodservice industry because consumers today have more choices than they ever had previously.

Through its use of brand identifiers, a foodservice operation can tell its unique story. For example, rather than simply describing its menu offerings, effective branding allows a foodservice operation to tell potential guests who it is and what it stands for. When properly done, it allows an operation to make an emotional connection with its target market. That connection allows all guests to feel good about the buying choices they make. This emotional connection can not only be useful in attracting new guests, but it can also grow long-term loyalty among current guests.

Key Term

Brand identifiers: The visible features that create a specific company brand. These can include any combination of name, logo, exterior building design, interior design, décor, music, menus, uniforms, pricing structure, services offered, and digital and traditional messaging.

Technology at Work

A foodservice operation's proprietary website continues to be one of its most important brand identifiers.

Fortunately, a large number of companies now specialize in designing restaurant websites. The best of these can ensure the website contains the operation's menu in an easy-to-read format, driving directions, online ordering information, payment options, and more.

Few foodservice operators have both the skills and tools necessary to create their own websites. This is especially so when the website must be designed to interface online ordering and payment options with the operation's POS system.

To review foodservice operation website designers and to see examples of their work, enter "restaurant web designers" in your favorite search engine and view the results.

Developing a Unique Brand

When developing their unique brands, foodservice operators may choose to create a **brand mark** or, in some cases, a **trademark**.

It is important for operators to ensure that all of the individual brand identifiers they utilize deliver a consistent message. For example, assume a foodservice operation emphasizes that its menu items are of excellent quality. If a guest enters the operation and encounters employees dressed sloppily in dirty and uncoordinated uniforms, a different message will be sent by the operation. When foodservice operators work hard to create a brand, they must work just as hard to manage that brand.

Consistency and brand management cut across many areas of foodservice operation. To cite just one example, if an operation desires to communicate

Key Term

Brand mark: The symbol or logo used to identify a foodservice operation.

The "Golden Arches" used in McDonald's advertising are a good example of a foodservice company's well-known brand mark.

Key Term

Trademark: A brandmark that has been given legal protection and is restricted for exclusive use by the owner of the trademark.

its **fine dining restaurant** status, then everything from the décor, the tableware, glassware, and the appearance of employees must support that upscale status.

It is easy to see that branding is one reason for the popularity of franchising in the foodservice industry. A **franchised operation** takes advantage of all of the branding work that has previously been undertaken on behalf of the brand. When customers see those familiar brand marks, they can more easily understand what they will experience as a customer.

It is easy to understand that a foodservice operator who simply wants to communicate that it serves "good food" will have a difficult time growing its customer base when many competitors are communicating so much more information to their target markets.

One excellent example of the power of branding can be seen in those colleges and universities who are very able to prepare high-quality hamburgers or pizzas. However, they know their student populations are much more likely to purchase these products from a franchised operation. Therefore, many of these educational institutions choose to host franchisees of these popular brands on their campuses.

Key Term

Fine dining (restaurant):
A restaurant experience that is typically more sophisticated, unique, and expensive than one would find in the average restaurant. Also known as an upper-scale restaurant.

Key Term

Franchised operation:
A business in which a foodservice operator (the franchisee) is allowed to use the brand identifiers of an organization (the franchisor) in exchange for the payment of franchise fees.

Find Out More

Franchise foodservice operations have tremendous advantages in communicating with potential guests. While the products they serve are important, in many cases it is just as important that the branding efforts they have undertaken make a positive impact on potential guests.

Some large franchisors offer multiple brands to their potential franchisees. For example, Louisville Kentucky-based YUM Brands operates the Kentucky Fried Chicken (KFC), Pizza Hut, and Taco Bells brands, among others.

To see the current size of the franchised foodservice market and to learn about its most popular franchisors, enter "largest foodservice chains" in your favorite browser and view the results.

While foodservice operators can select from many brand identifiers as they define their brands, one extremely important identifier common to all operations is its name.

A foodservice operation's name must be easy to pronounce and easy to remember. In most cases, the use of foreign terms or letter combinations that are difficult to pronounce should be avoided. For many consumers, it will be the operation's name alone that first captures their attention. Figure 3.1 illustrates a few examples of how the use of "name" can be a major portion of an operation's marketing message even though it is used in many different ways.

Operation Name	Location	Message
Blaze Pizza	Pasadena, California-based chain	Fast preparation of made-to-order pizza
Ruth's Chris Steak House	Winter Park, Florida-based chain	Owner's first name and the main product featured
Like No Udder	Providence, Rhode Island	Vegan ice cream company with memorable phrasing
Tequila Mockingbird	Ocean City, Maryland	Fun-themed Mexican cantina using well-known pop culture reference
Life of Pie	Ottawa, Canada	Bakery and pie shop playing off the name of the popular book, *Life of Pi*, written by Canadian author Yann Martell
The Notorious P.I.G.	Missoula, Montana	Barbeque house playing off the name of the iconic American rapper
The Codfather	Reno, Nevada	Fun-themed seafood shop using pop culture reference
The French Laundry	Napa, California	Fine dining restaurant named for the original purpose of its building
Bubba Gump Shrimp Company	Division of Houston, Texas-based Landry's Restaurants	Seafood restaurant playing off the name recognition of the fictional book and movie character Forrest Gump
Kale Me Crazy	Atlanta, Georgia-based chain	Salad bar restaurant featuring fresh, organic, and raw foods
The View Lounge	San Francisco, California	Bar named after its unique location overlooking the city
Mary Mahoney's	Biloxi, Mississippi	Owner's first name and last name only

Figure 3.1 Restaurant Name Examples

Product-Focused Branding

While an operation's name is of key importance and must be carefully chosen, it is only one of several brand identifiers that can be used to communicate with an operation's guests. In many cases, the menu items offered will feature heavily in an operation's target market communications.

For example, guests seeking Italian food will be looking for operations whose communications feature Italian-style menu items. Taken another step, an operation may communicate that it features "Northern-style Italian" or "Southern-style Italian" because knowledgeable "Italian food-seeking" diners likely understand the distinction between these two cuisines. Similarly, Asian-style foodservice operations will likely want to emphasize that they feature Vietnamese, Korean, Thai, or Chinese-style menu items as these different cuisines may appeal to different guests.

In all cases, when an operation communicates product-focused information to its target market, the messages should clearly identify the type of menu items that are offered.

Service-Focused Branding

In some cases, a foodservice operation may focus its marketing message on the service style offered to guests, rather than its individual menu items or cuisine. In fact, service style is so important to guests that the foodservice industry classifies operations based primarily on that operating characteristic. While there is minor variation among industry analysts, Figure 3.2 lists the most common service-related classifications of food operations.

There are additional and unique service styles of operations available in many foodservice markets. These include:

Family style: In these operations, food is served on large platters for guests to share. The targets market for these types of operations are large groups and families.

Buffet style: In these operations, food items are presented on serving tables or heated steam tables and guests serve themselves. In most cases these styles of operation are marketed as "all-you-care-to-eat" and often appeal to value-oriented diners because one price is charged regardless of the amount of food selected.

Food trucks and concession stands: These operations typically serve a smaller menu of a singular type of food or cuisine (e.g., burgers, hot dogs, and other sandwich types, or Mexican, Asian, and the like). Some food trucks vary their locations daily, and other operations are typically outdoors or at sporting events, festivals, and fairs. Menu prices range from low to moderate.

Classification	Common Service-Related Characteristics
Fine Dining	Formal dress code for staff
	Use of waitstaff for "at table" menu item delivery
	High-end décor and table/dishware
	Menus feature exotic or complex dishes
	Wide selection of wines
	Relatively high-priced
Casual	Informal dress code for staff
	Use of waitstaff for "at table" menu item delivery
	Low-key (casual) atmosphere and unique décor
	Moderately-priced
Fast Casual	Informal dress code for staff
	Counter service of made-to-order foods picked up by guests
	No alcohol service (in most cases)
	Moderately-priced
Quick Service	Informal dress code for staff
	Counter service food picked up by guests
	Take out and drive-through (in most cases) and menu delivery options are available
	No alcohol service (in most cases)
	Low-priced

Figure 3.2   Common Service Classifications of Foodservice Operations

Ghost restaurants: Because these operations have no brick-and-mortar location they do not allow for in-person dining. Instead, these operations have a very high-profile website and online media presence. The menu items sold by these operations are either picked up by guests or delivered to them. Menu prices typically range from low to moderate, but they can be high-priced if the restaurant is operated by a well-known chef or locally well-known foodservice operator.

Regardless of whether an operator elects to feature product or service style, operators will face unique challenges and opportunities when making their brand identity decisions and crafting their marketing messages.

Factors Affecting Product-Related Messages

Nearly all foodservice operators are justifiably proud of the menu items they offer for sale. As a result, their tendency is to make the products offered for sale a major focus of their operations' marketing message and branding efforts.

Product-related marketing messages make good sense when an operation is confident it can consistently deliver what is promised on its menu. For example, if an operator desires to offer fresh salads, then those salads must consistently be delivered exactly according to the standards the operator establishes. It is poor marketing to emphasize an area of a food operation that could disappoint, rather than delight, guests.

When operators seek to use the products sold in their operations as brand identifiers, there are a number of different factors and focus areas that can be utilized. The most important of these include:

Product Quality
Preparation Method
Serving Size
Price
Convenience
Location

Product Quality

Those operators who emphasize product quality in their marketing messages understand that the freshness of ingredients is a significant factor affecting their customers' satisfaction levels. Cooking styles and seasonings are important, but experienced food production professionals know that quality end-products begin with quality ingredients.

For example, emphasizing the use of United States Department of Agriculture (USDA)–Graded Choice or Prime beef steaks communicates a high-quality level to customers. In other cases, emphasizing easily recognizable branded ingredients (examples: the use of A.1. Steak Sauce, Heinz catsup, Oreo cookies, or Jack Daniel's Barbeque Sauce) helps communicate high quality. Taco Bell has also had great success emphasizing the quality of their tacos by co-branding them with "Doritos" brand corn chips.

Emphasizing the use of locally sourced and seasonal ingredients is another common way to communicate product quality. An emphasis on "good-for-you" ingredients that are natural, low in sugar, cholesterol, and sodium, and high in antioxidants can also highlight product quality.

It is important to note that quality is not the same as terms such as "healthy" or "gourmet." Assume a customer sees a menu photo of delicious-looking pancakes with a side of crisp bacon and a container of fresh maple syrup and orders these items. If the order presented to the guest contains these items properly cooked, high-quality food was prepared and served. If, however, the bacon is burnt, the pancakes are flat, and the syrup is too thin, the menu items will disappoint the guest because quality products were not received.

It is also essential to recognize that the utilization of product quality as a brand identifier is different from using the same terms when writing menu items' descriptions. Proper development of **menu copy** is important, and the use of the menu as a marketing tool will be addressed in detail in Chapter 7. The use of product-related terms as brand identifiers, however, encourages customers to select a particular operation for the first time, and then the customer can learn about the operation's menu and menu item copy as they make their selections.

Key Term

Menu copy: The words and phrases used to name and describe items listed on a foodservice menu.

When deciding to emphasize their product quality, operators must be especially aware of the importance of consistency: the delivery of what was promised every time the item is selected. When high quality is emphasized, it is assumed by guests that high-quality products will be delivered time after time.

In many communities, local websites, newspapers, and/or radio will often seek consumer input to rate nearby restaurants and then publicize the results. These "most popular" voting systems often classify restaurants by type. For example, operations can be classified as the most popular (favorite) pizza, steak, food truck, or breakfast restaurant in a city or area. When a foodservice operation scores well on these quality-related popularity contests, the results can be used to promote the restaurant's quality.

Find Out More

Foodservice operators can, of course, make their own proclamations touting the quality of their menu items. When outsiders not affiliated with the operation do the same, however, it can make a big difference in the eyes of many diners.

The most famous of the outside review groups in the foodservice industry is Michelin. Their Michelin Guide rates foodservice operations on a scale of 1–3 stars: with 3 stars being their highest rating. It is incredibly difficult to gain even one star on the Michelin rating system. Worldwide only 135 operations rated three stars in 2021.

In the US, an increasingly popular rating system is that of the James Beard Foundation. Named after the famous chef, foodservice operations whose chefs and businesses are nominated for a James Beard award immediately gain national attention, and their nominations are widely publicized by the selected operations as an indicator of their quality.

To learn more about James Beard nominations as well as the new "American Classics" category that honors family-owned restaurants across the country, enter "James Beard Award nomination criteria" in your favorite browser and view the results.

Foodservice operations that emphasize quality as a major brand identifier know that product quality is usually identified as the most important characteristic associated with the customer returning to the business. Other brand identifiers may encourage customers to try an operation for the first time, but the quality of food brings them back time and again.

Key words commonly used in quality-focused product marketing include terms such as Voted # 1, Award Winning, Most Popular, Reader's Choice, Chef-made, Seasonal, Fresh-baked, Free-range, Fresh-picked, Organic, Low-fat, and Sugar-free.

Preparation Method

In many cases, an operation's unique preparation style can set it apart from its competitors. For example, the pizza operator who emphasizes that its pizzas are "wood-fired" communicates that its pizzas are unique and superior because of their preparation method.

The use of terms such as Cajun-style, Slow-roasted, Hibachi-grilled (Japanese items), Southern-fried (chicken), Flamed-grilled, and "House-made dressings" are several examples of using preparation style as a major component of marketing the uniqueness of menu items.

Additional key words commonly used in preparation method-focused marketing include Poached, Rotisserie, Broasted, Sautéed, **Sous vide** (pronounced "soo VEED") Barbecued, Old-fashioned, Slow-roasted, and Slow-cooked.

When an operation's menu items are very similar to those of its competitors, an emphasis on preparation style may be a good way for the operation to use the preparation (cooking) method to differentiate itself from its competitors.

Key Term

Sous vide: A method of cooking food that has been vacuum-sealed in a plastic bag and immersed in a regulated, low-temperature water bath.

Serving Size

Serving size has for many years been a very traditional way for foodservice operators to differentiate themselves from their competitors. One example: in 1957, the Burger King Corporation first began to market a large hamburger called the "Whopper." At that time, its major competitor (McDonald's) was still selling only small (1.6-ounce patty) hamburgers. Burger King then became known in the eyes of consumers as the "Home of the Whopper."

It would not be until 1973 that McDonald's responded to this market challenge by introducing the "Quarter Pounder" for customers wanting a larger hamburger patty than their traditional burger. In both examples, the companies wanted to communicate a brand identifier that large hamburger serving sizes were available.

In Chapter 1, "Value" was defined as "the amount paid for a product or service compared to the buyer's view of what they receive in return." Some foodservice customers view the concept of "large quantity" as equal to the concept of "good value." The average foodservice operator also experiences more complaints from guests who feel their portion sizes were too small for the price paid than those believing portion sizes were too large. Therefore, some operations who utilize portion sizes as a brand identifier are doing so for the right reasons.

Find Out More

The promotion of products based on their portion sizes must be approached carefully. This is especially so when guests can consume as much as they want when paying one price.

In 2003, Edna Morris, the President of the Red Lobster chain, announced the introduction of the company's "Endless Crab" promotion. For $23.99 guests could eat all of the snow crab they desired. Morris certainly must have felt it was a good idea because it was widely advertised.

However, the plan was not a good idea! "It wasn't the second helping, it was the third that hurt," Joe Lee, former chief executive, explained to investors. "And the fourth," agreed Dick Rivera, Morris' predecessor, and successor to the president's position at Red Lobster (as quoted by *USA Today*).

As it turns out, providing an all-you-can-eat option on some menu items can cause disastrous financial results. To learn more about the specific problems Red Lobster encountered after making this quantity-focused branding decision, enter "Red Lobster crab promotion fiasco" in your favorite browser and view the results.

Using portion size as a brand identifier will no doubt continue. The Olive Garden chain's promotion of its "Never Ending" soup, salad, and breadsticks, and Paula Dean's Family Kitchen's promotion of "Unlimited Refills" are two current examples.

Key words commonly used as serving size-focused brand identifiers include Jumbo, Giant, Extra Large, Never-ending, All-You-Can-Eat, Endless, and Unlimited.

Price

There is no doubt that price is a significant part of the marketing efforts of many foodservice operations. In fact, the effective use of price in marketing is so important that it will be addressed in detail in Chapter 6. Price is also used by many foodservice operators as a brand identifier.

The proper use of price as a brand identifier is often more complex than it may appear. To best understand how price can properly be used as a major brand identifier it is important to recognize that all consumers like low prices. However, this is only true when the consumer associates low price with good value. For example, if consumers associate "low price" with cheap, inferior, or low-quality products, then the low price can create a negative impression in the eyes of the consumer.

Perhaps the most significant use of price as a brand identifier is illustrated by many foodservice operations' implementation of a **value menu**.

In 1989, Wendy's introduced its 99¢ Value menu. McDonald's created its Dollar Menu in 2002. Burger King rolled out its similarly priced BK Value Menu in 2006. These chains, and others, still offer a variety of value menu items at the same 99¢ to $1.00 price point. The price points used in some value menus today are higher; often $1.99, $2.99 or $3.99.

Key Term

Value menu: A group of menu items designed to be the least expensive items available. The portion size and number of items included in a value menu differ by operation, but all are designed to attract consumers with low-priced menu items.

Some customers love value menus, as do franchise companies (whose franchise fees are based on a franchisee's gross sales, not on the franchisee's profits). However, many franchised operators are less than enthusiastic about the use of low prices to drive unit sales.

Some industry experts also have mixed feelings about the true "value" of value meals because they are seen as a mixed blessing. Advocates state that value menus drive customers to the stores. They maintain that customers like value menus and, the reasoning continues, customers must be given what they want.

Those who advocate against value menus argue that ever-increasing labor, food, and other operating costs negatively affect the opportunity to generate reasonable profits when value menu sales are emphasized.

Rather than offer actual value menus, many foodservice operations utilize low prices selectively. For example, an operation may offer a reduced price on a specific menu item on a certain day, or at a certain time, as a brand identifier.

Another common use of price as a brand identifier relates to both price and quantity purchased. For example, a pizza operator may offer a lower "per-pizza" price when two rather than one pizza (s) are purchased. For example, if one medium pizza sells for $19.99, the purchase of two pizzas for $29.99 is designed to increase the customer's perception of the value received for the price paid. This use of low price may also encourage additional purchases that, in turn, increase the operator's average sale per order and its per-sale profits.

Find Out More

Faced with ever-increasing food and labor costs, restaurant chains continually revise their value meals to strike a balance between increasing guest traffic counts and maintaining acceptable profit margins. The efforts to do this vary by chain and are constantly evolving. Some operations selectively raise prices while others create products specifically for sale on their value menus.

To remain current on the various ways foodservice operators are modifying their value menus and to review potential ideas for use in your own operation, enter "changes to value menus" in your favorite browser and view the results.

Price will likely remain a significant brand identifier for many foodservice operations. Key words commonly used as price-related brand identifiers include Value Menu, Value Meal, Buy One Get One (BOGO), Two-for-One, Special, Early Bird, Happy Hour, and Limited Offer.

Convenience

For many guests convenience is a primary factor in helping to determine the foodservice operation they want to visit. Consider the office worker who has ½ hour for lunch. In this example, the worker must find a foodservice location, place their order, receive it, and still have time to eat before returning to work on time.

Convenience can be described in different ways by different customers. For some customers, convenience means being able to easily pick up an order on their way to work or home. For others, it may mean an operation that is within easy walking or biking distance.

When a foodservice operation decides to utilize convenience as a brand identifier, it is most often promising its customers ease of purchase, order accuracy, and speed of delivery. To achieve these promises today, an effective **online ordering system** is increasingly a necessity.

An online ordering system accomplishes several objectives. First, it clearly displays the menu items offered for sale. This may be done with menu copy and item descriptions, but the best online ordering systems also include photographs of the menu items being sold. Second, it must make menu item selection easy and intuitive for users. Finally, popular payment options should be offered. These can include credit and debit card acceptance and the use of other increasingly popular online payment forms (mobile wallets) such as PayPal, Google Pay, Apple Pay, and others.

Key Term

Online ordering system: A software program that allows a foodservice operation to accept and manage orders placed via a website or a mobile app.

Technology at Work

Demand for online ordering grows among foodservice customers. This has prompted operators to find solutions that enable customers to easily browse menus, make selections, and pay for their pick-up or delivery orders online. Fortunately, many currently offer online ordering systems that provide all these solutions.

Convenience in online ordering essentially means that when a guest orders food, the entire process is streamlined and seamless. For example, customers can order on numerous channels including websites, mobile apps, and social media. Available systems offer ease of use and flexibility to change orders after they are placed and provide multiple payment options without requiring customers to sign up or create a separate account.

To review some companies currently offering online ordering systems and their costs, enter "best online ordering systems for restaurants" in your favorite search engine and view the results.

The increased use of drive-through and pick-up service options has prompted many foodservice operators to use on-premises directional signage to ensure customer convenience. They know that, even if an order is perfectly prepared and packaged, low satisfaction levels will result if arriving guests don't know where to pick up their items or if the lines to do so are too long. Food operations that make convenience part of their marketing message must ensure that they address prompt pick up of their products along with the preparation of the products themselves.

Convenience is a significant brand identifier for many foodservice operations. Key words commonly used as convenience-related brand identifiers include Fast, Speedy, Express, Quick, Ample Parking, and Easy-In Easy-Out.

Location

It has long been an old adage that the three most important factors predicting the success of a commercial foodservice operation are location, location, and location! It is true that location is an extremely important part of planning and operating a foodservice. An operation's location may also, in some cases, be used as a significant brand identifier.

For example, consider the foodservice operation whose window placement allows guests to have panoramic views of the ocean while dining. Regardless of the type of cuisine it offers or its service style, this operator may well determine that promoting its ocean view is a significant way to attract guests. Similarly, a foodservice operation located near a major **attraction** may promote that nearness in their marketing efforts.

In some cases, a foodservice operation can communicate a great deal about itself simply by stating where it is located. For example, an oper-

Key Term

Attraction: A place or event that draws visitors by providing something of interest or pleasure.

ation that states that it is located "In the heart of scenic downtown" is communicating its environment to locals and tourists alike.

Experienced foodservice operators know that guests put a premium on location convenience. Therefore, developing a marketing message targeting those who live near an operation often makes good sense. If an operation advertises a pick-up service, it is likely that those who order from the restaurant will live in the area because few customers want to drive long distances to pick up their food orders. Advertising that reminds guests of its nearby location is one way for an operation to target those living in the general vicinity.

In some cases, such as for those operating food trucks, their operation's location may change regularly. For example, the food truck may be located in one area for its lunch service and move to another area for dinner service. In these situations, the food truck operator requires an effective and efficient method of communicating their current and future locations to potential guests.

Technology at Work

One popular method of utilizing location to promote a foodservice operation is through geo-targeting. Geo-targeting refers to the practice of delivering different content or advertisements to consumers based on their geographic locations. Other terms for Geo-targeting include geolocating, geofencing, or proximity marketing.

One good example of Geo-targeting is Google search results. If a user utilizes a smart device to search for "coffee shops," Google will utilize location data based on the IP address of the smart device to supply only information about coffee shops that are close to the searcher's location.

Facebook, Instagram, Snapchat, Twitter, TikTok, Pinterest, and LinkedIn currently have programs for helping to identify custom audiences.

Not surprisingly, there are advertising-related vendors who specialize in using geo-targeting results to fine-tune a foodservice operation's marketing message. To review some of these companies and how they can assist in reaching nearby customers, enter "location-based marketing tools for restaurants" in your favorite search engine and review the results.

Key words commonly used as location-related brand identifiers include Secluded, Lake View, Ocean View, Mountain View, Overlooking, Panoramic, Sunset, Located in the heart of, Located just steps from, and Located near.

What Would You Do? 3.1

"How were our sales last month? asked Frank Ginzner, the owner of the Big Steer Steakhouse.

Frank was talking to Carlos Magana, the manager of the Big Steer.

"Dinners were great," replied Carlos, "But lunches are still a little slow."

"Well," said Frank, "That's pretty much always been the case for us since we opened. When you serve high-end steaks and wines like we do, people tend to choose us for special occasions and their evening meals. They like the leisurely pace of our service. But at lunchtime, not so much."

Well, we need to fix that," said Carlos, "We need to convince people that this is a great place for lunch."

Assume you were Carlos. Do you think the target market for the dinner business and lunch business at the Big Steer Steakhouse is the same? How would you go about creating and selecting product-related brand identifiers for the Big Steer Steakhouse that would help it create a "lunch-focused" marketing message?

Service Offering Messages by Type of Operation

While a focus on products served will no doubt continue to be important to many foodservice operators, service levels provided may also be important when developing the marketing message.

One good way to review the options available to operators wanting to focus on their service style of delivery is to utilize the common classifications of foodservice operations. As previously identified, these common classifications include:

✓ Non-commercial Foodservice Operations
✓ Quick Service Restaurants (QSRs)
✓ Fast Casual Restaurants
✓ Casual Service Restaurants
✓ Fine Dining/Upscale Restaurants
✓ Ghost Operations

In addition to these traditional commercial for-profit classifications, many foodservice professionals operate their units in the non-commercial market.

Non-Commercial Operations

Non-Commercial foodservice operations are those in which serving food and beverages is secondary to the goals of a larger "host" organization. For example, the primary purpose of a college or university is not to serve food and beverages. However, to achieve their primary purpose of educating students, food and beverage services are often provided for students and staff at these institutions. In many cases, non-commercial foodservice operations are self-operated by the host organization. In other cases, companies that specialize in operating non-commercial operations are retained to do so by the host organization.

There are a number of names commonly used to identify non-commercial operations. These include non-profit, institutional, contract feeding, on-site foodservices, or managed services. The number of foodservice operations that are non-commercial is vast and varied. These include food operations in:

K–12 education
Colleges
Universities
Hospitals
Nursing homes
Retirement centers
Correctional facilities
Military facilities
Businesses
Factories
Private clubs
Sporting arenas and stadiums
National and State parks
Transportation providers (e.g., airlines, trains, and cruise ships)

Non-Commercial foodservice operations can include those which are classified as quick service, fast casual, casual, and fine dining/upscale. While the success of a commercial (for-profit) foodservice operation is most often determined by its sales volume (dollars), the success of non-commercial operations is most often evaluated by participation rate or the number of people it serves. Thus, a clear and effective marketing message is just as important to non-commercial operations as it is to commercial operations.

Quick Service Restaurants (QSRs)

QSRs offer food items that require minimal preparation time and that are delivered shortly after they are ordered. The typical QSR may offer guests limited indoor seating, and most have drive-through lanes. Guests' orders may also be prepared and packaged for pick-up, or for delivery to the guest's preferred location.

In many ways, regardless of their menu offerings, QSRs compete among themselves for the same target markets. A potential guest considering the various QSRs available to them may choose the operation they want to frequent based on product, but they may just as frequently base it on an operation's location and convenience.

Key words commonly used as service-related QSR brand identifiers include Open 24 hours, Late-night, All-night, Drive-through, Fresh, Fast, Hot, and Value.

Find Out More

Operators of QSRs may seek to focus on the product served, or their service style, as they develop and deliver marketing messages. To review how various operators do this, enter the name of one or two well-known QSR operations in your favorite browser and view the results.

During your review pay particular attention to whether they are featuring their products, their service style, or both while delivering their marketing messages.

Fast Casual Restaurants

Fast casual is the most recent of the restaurant classifications. The concept originated in the United States in the early 1990s but did not become common until the beginning of the 2010s.

Currently, popular fast casual brands include Chipotle Mexican Grill, Panera Bread, Jersey Mike's Subs, Shake Shack, and Panda Express. Some industry analysts would argue that the origin of the fast casual restaurant concept began in the 1970s, when the Burger King restaurant chain utilized, as a major brand identifier, the slogan "Have it Your Way," based on the chain's willingness to tailor hamburger orders to each individual guest's tastes and preferences.

The typical fast casual operation does not offer at-table service, but rather it will prepare menu items to their customers' individual preferences. For example, a fast casual Mexican-style food operation allows guests to make ingredient selections as they pass through a cafeteria-style line and select specific ingredients for their tacos, burritos, and/or food bowls.

In their marketing messages, fast casual restaurants typically address the ability of guests to customize their menu choices and the freshness of their ingredients. Most fast casual restaurants also offer their menu items for pick-up or delivery, and this is often a significant part of their marketing messages.

Key words commonly used as service-related brand identifiers in fast casual operations include Fresh, Made-to-order, Custom, Fast, and Fun.

Find Out More

Operators of fast casual operations may focus on products served or service style as they deliver marketing messages. To review how various operators do so, enter the name of one or two well-known fast casual operations in your favorite browser and view the results.

During your review pay particular attention to whether they are featuring their products, their service style, or both, as they deliver their marketing messages.

Casual Service Restaurants

Casual service restaurants are those that typically provide service at a guest's table. Also known as a "sit-down" operation, guests seated by an employee make their menu selections, which are then delivered to their tables. A casual service restaurant may serve any type of cuisine. Pricing in casual service restaurants is typically higher than at QSRs and fast casual operations, but lower than those charged in fine dining operations.

Currently, popular casual service restaurant chains in the United States and Canada include Applebee's, Olive Garden, Red Lobster, Texas Roadhouse, The Cheesecake Factory, and PF Chang's China Bistro. The casual service restaurant segment also includes many unfranchised independent operations.

Casual operations are popular with guests because they offer a personal experience when eating out, and they charge affordable prices. Casual operations are especially popular for families who want a dining out experience. The lower price point of casual service restaurants makes it easy for younger and lower income people to dine out as well.

While casual service restaurants emphasize their onsite dining style in response to consumer demand, they increasingly offer their menu items for pick-up or delivery.

Key words commonly used as service-related brand identifiers in casual operations include Outdoor, Patio, Clean, Comfortable, Friendly service, Quick, Prompt, Authentic, Family-friendly, Tables, and Booths.

Find Out More

With so much emphasis on new forms of foodservice operations, it can be difficult for traditional dine-in casual service restaurants to distinguish themselves from their competition. But creative operators can do just that. Consider Lambert's Cafe (Foley, Alabama and Sikeston, Missouri).

Anyone familiar with this high-quality restaurant is also familiar with its focused marketing message. Lambert's bills itself as the "Home of the Throwed Rolls."

And that is exactly what you will see if you visit Lambert's. Rolls, hot from the oven, are thrown across the dining room for eager diners to catch. It is fun for guests, and it is memorable.

To see a brief video capturing this activity, go to YouTube, enter "Lambert's Home of the Throwed Rolls," and see how they do it.

Fine Dining/Upscale Restaurants

The food and beverage items served at fine dining restaurants will be of the highest quality and made with the finest ingredients available. Service levels will be exceptional. In most cases, the interiors of the operations will be unique and beautiful. Flatware, dishware, glassware, and linens will be of the highest quality as well.

Consumer expectations in dining restaurants are very high. However, it is important to recognize that fewer customers will visit a fine dining property than a casual service restaurant on a daily basis. In more cases, fine dining establishments are chosen for special occasions, celebrations, and business-related meals.

While it is relatively uncommon, some upscale restaurants do offer their menu items for pick-up or delivery, and this service method is increasing.

Key words commonly used as service-related brand identifiers in fine dining operations include Chef, Finest, Charming, Gourmet, Famous, World-class, Butcher, Prime, Authentic, Seasonal, Stars, Sustainably Sourced, Elegant, Cozy, Formal, Intimate, and Romantic.

Find Out More

Upscale does not always mean formal.

You are likely familiar with Ruth's Chris Steakhouses, the upscale USDA Prime steakhouse founded by Ruth Fertel in New Orleans, Louisiana, in 1965. The product quality, ambiance, and service levels provided allow Ruth's Chris to charge premium prices and deliver excellent value.

You may not be as familiar with the equally pricy but extremely casual Zingerman's Deli (Ann Arbor, Michigan), where commonly-sold items include

English farmhouse cedar cheese ($45.00 per pound), Kentucky smoked break-
fast sausage ($17.00 per pound), and deli sandwiches sold at prices above
those of Manhattan's most popular 7th Avenue delicatessens.

Prices at Zingerman's reflect their carefully crafted, but very laid-back
image. That image consists of a commitment to providing a wide variety of
extremely high-quality food products, exceptional customer service in a very
relaxed atmosphere, and as a result, outstanding value.

To see this example of how casual and upscale can be blended, enter
"Zingerman's Deli Ann Arbor" in your favorite browser and view the results.

Ghost Operations

Ghost kitchens (see Chapter 1) face unique challenges in marketing their operations. The reason for this is that they typically have only a little, or no, exterior
signage. As a result, they must deliver their marketing messages through digital
means or through print and/or broadcast advertising.

Effective digital marketing for most ghost operations begins with a vibrant website that shows pictures and/or videos of the foods being offered. Navigating the
website must be easy, and ordering must be simple. Pick-up and delivery options
must be made very clear. In most cases a ghost operation, regardless of its classification, will also have a proactive e-mail campaign, with weekly e-mails directed to
past guests. Other common forms of communication used by ghost operations
include Twitter, Pinterest, Facebook, and other social media platforms.

A foodservice operation's brand identification always begins with an operation's carefully selected name. Then, decisions must be made on whether to focus
on products or services as major brand identifiers. Doing these things effectively
will help create a memorable marketing mes-
sage for an operation's target market.

Regardless of the brand identifiers utilized by
foodservice operators and the marketing messages directed to their target markets, all operators must be able to *deliver* that message in a way
that maximizes its impact. Doing this requires an
understanding of available **marketing channels**
and a keen understanding of buyer behavior, and
these are major topics in the next chapter.

Key Term

Marketing channel: A method
or form of delivering a
marketing message to an
operation's current and
potential guests. Also
commonly known as a
"communication channel."

What Would You Do? 3.2

"The space and the equipment are perfect! said Max. He and his business partner Sonya had just finished inspecting some production space that was available for lease in a building shared by three other ghost food operations. Max and Sonya were opening "Soupy Sales," a delivery-only ghost restaurant specializing in the sale of up-scale soups, chilies, and stews.

"OK," said Sonia. "If you think the facility will meet our needs, I am all in with it! It's exciting!"

"I can have the kitchen up and running in three weeks," said Max.

"So now the question is, "How do we let our customers know what we are selling and why they will love us!" said Sonya.

Assume you were asked to assist Max and Sonya in the initial marketing of Soupy Sales. As a ghost restaurant, how important do you think it would be for them to select the right brand identifiers as they craft their initial marketing messages to their target markets? How would you advise them to do it?

Key Terms

Brand	Franchised operation	Attraction
Brand identifiers	Menu copy	Non-Commercial
Brand mark	Sous vide	(foodservice operation)
Trademark	Value menu	Marketing channel
Fine dining (restaurant)	Online ordering system	

Operator's 10-Point Tactics for Success Checklist

Evaluate your need for, and the current status of, each of the following operational tactics. For those tactics you think are important, but not yet in place, develop an action plan for its implementation including who will be responsible for the tactic's completion and the target date by which it should be completed.

Tactic	Don't Agree (Not Done)	Agree (Done)	Agree (Not Done)	If Not Done Who Is Responsible?	Target Completion Date
1) Operator understands the importance of creating a unifying marketing message to differentiate their operation from their competitors' operations.	——	——	——		

Tactic	Don't Agree (Not Done)	Agree (Done)	Agree (Not Done)	If Not Done Who Is Responsible?	If Not Done Target Completion Date
2) Operator has carefully considered the importance of utilizing brand identifiers in their marketing message.	——	——	——		
3) Operator has assessed the use of "Product Quality" as a potential factor when selecting brand identifiers.	——	——	——		
4) Operator has assessed the use of "Serving Size" as a potential factor when selecting brand identifiers.	——	——	——		
5) Operator has assessed the use of "Price" as a potential factor when selecting brand identifiers.	——	——	——		
6) Operator has assessed the use of "Convenience" as a potential factor when selecting brand identifiers.	——	——	——		
7) Operator has assessed the use of "Location" as a potential factor when selecting brand identifiers.	——	——	——		
8) Operator has chosen the primary product-related factors to be used in crafting the marketing message for their own specific operation.	——	——	——		
9) Operator has identified their type of operation as a first step in creating a service-offering-related marketing message.	——	——	——		
10) Operator has reviewed and selected for use the most important brand identifiers associated with their own particular type of operation.	——	——	——		

4

Delivering the Marketing Message

What You Will Learn

1) The Importance of Assessing Marketing Channels
2) The Four Classifications of Buyers
3) How Buyers Make Purchase Decisions
4) The Legal, Ethical, and Social Responsibilities of Marketing

Operator's Brief

In this chapter, you will learn that all foodservice operators must decide how to deliver their marketing messages to current and potential guests. To do that well, a foodservice operator must consider traditional and digital alternatives and then choose those that will be the most cost-effective and make the biggest impact on guests.

As they evaluate the best ways to deliver their marketing messages, operators must pay particularly close attention to any new communication channels their guests are using. By doing so they can help ensure their messages will reach the largest possible target audiences and their marketing dollars will be well spent. This is true whether the operator is utilizing traditional and/or web-based communication channels.

Regardless of the communication tools they use, all foodservice operators must understand those to whom they sell. Experienced foodservice managers know that there are four distinct types of buyers in the foodservice industry, and each has a slightly different view of purchasing. In this chapter, you will learn about these four types of buyers, as well as the mental process all buyers go through as they make their purchase decisions.

Foodservice professionals make many decisions as they choose their communication tools and address the motivations of the guests who will buy from them. In general, foodservice operators have a great deal of freedom as they craft and deliver their marketing messages. However, in this chapter, you will learn that there are legal, ethical, and social responsibilities that also impact how operators can and should market their products and services.

CHAPTER OUTLINE
Assessing Marketing Channels
Marketing Channel Options
Assessing New Communication Channels
Understanding Buyer Behavior
The Four Buyer Types
Buyer Behavior
Operator Responsibilities in Target Market Messaging
Legal Responsibilities
Ethical Responsibilities
Social Responsibilities

Assessing Marketing Channels

In the previous chapter a marketing channel (also known as a communication channel), was defined as a method or form of delivering a marketing message to a food operation's current and potential guests.

Today's foodservice operators have an incredible number of alternative marketing channels available to them. With so many alternatives to choose from, it is essential that operators carefully assess their available options. Doing these analyses properly involves first understanding the purpose and costs of various marketing channels and then identifying those that fit best with the message the operation is trying to send to its target market.

Marketing channels continue to evolve in the hospitality industry. As a result, foodservice operators must continually assess new and emerging communication channels to identify those that may be most valuable.

Marketing Channel Options

In years, past foodservice operators were primarily required to use one-way communication when interacting with their guests. For example, years ago, a foodservice operator might place an advertisement in the local newspaper's dining section.

As shown in Figure 4.1, while such advertising may or may not have been effective, the communication was only from the operator to potential guests. There was no option for any type of immediate guest feedback.

Today, an increasing number of communication tools allow guests to respond to an operator's marketing messages, and often in real time. For example, a foodservice operator today might send out an e-mail advertisement to guests

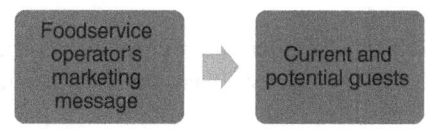

Figure 4.1 One-Way Marketing Communication

containing a return feature button that allows guests to indicate the day and time they want to make a dinner reservation.

As shown in Figure 4.2, this method of two-way communication allows operators to talk *with* their guests, rather than merely talk *at* their guests.

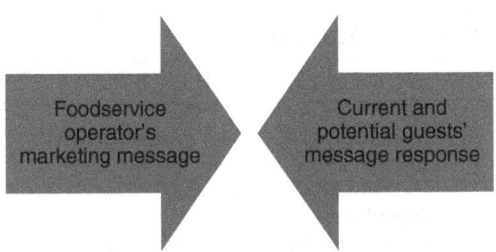

Figure 4.2   Two-Way Marketing Communication

There are several effective ways to assess the one-way and two-way communication tools available to foodservice operators. One good way to do so is to classify them as either a traditional communication channel or as a **web-based communication channel**.

Whether traditional or web-based, different communication channels have different strengths and weaknesses. For example, radio advertisements can reach a broad audience and be very cost-effective. However, that same communication channel cannot show pictures of menu items or interiors of restaurants that might be very appealing to guests.

Key Term

Web-based communication channel: A communication method that relies on the internet and/or a smart device for its message delivery.

Also referred to by some marketing experts as a "Digital" or "Online" communication channel.

Alternatively, television advertising can show close-ups of menu items, beautiful restaurant interiors, and guests enjoying themselves while dining. However, television advertisements can be costly to produce and reserving prime advertising times for running television ads can be expensive.

Experienced foodservice operators know that they will most often utilize more than one communication channel as they speak to their guests. As a result, it is important that operators choose the best traditional or web-based communication channel to deliver their intended messages.

Traditional Communication Channels
There are numerous traditional communication channels. One effective way to assess them is to first classify them as either a print channel or a broadcast channel.

Print Channels
Print is a communication channel that foodservice industry operators have relied on heavily in the past. Today it is still a popular and effective one-way communication

channel. The types of printed communication pieces used by operators can be extensive and include:

✓ Billboards	✓ On premise menus	✓ Direct mail
✓ Exterior signage	✓ Take away menus	✓ Coupons
✓ Interior signage	✓ Newspapers	
✓ On table signage	✓ Magazines	

Operators selecting the use of print as a one-way communication channel must understand some basic facts about the channel.

1) The print piece must be easy to read

Whether the print piece is a billboard seen on a highway, or a one-page flyer to be handed out to guests on-site, the type size and style used should be large enough and clear enough for guests to easily read the ad. The layout of the printed material should also appear uncluttered and in balance.

2) If possible, the print piece should include pictures

Adding graphics to a print advertisement increases the chances that a reader will actually read it. Graphics included in print ads may be photographs, drawings, or maps. Especially in the foodservice industry, photos of food items, diners, and attractive interior shots help draw the reader's attention. Also, it is important to remember that readers of print ads are more likely to remember the photos they have seen than the printed words they have read.

3) Ads produced in color draw more attention than black-and-white ads

In most cases, operators creating their print ads can produce them in full color or in black and white. Ads printed in full color are more expensive to produce but, in nearly all cases, the improvement in readability and effectiveness is worth the extra expense.

4) The print ad should include contact information

In most cases, print ads should include the name, address, and contact information (e.g., phone number or web address) for the business producing the ad. Most consumers are exposed to a large number of print ads every day, and it is not reasonable to assume that they will remember the details of each ad that has been read.

Logos, phrases, and other brand identifiers (see Chapter 3) can be used to make ads memorable. However, this will generate little value if the ad's reader does not know how to make a reservation, call in an order, or visit the operation in person.

5) The print piece must be grammatically correct

It is imperative that printed ads be carefully proofed and edited. Errors in spelling or grammar are never appropriate. In most cases, it is a good idea to have several individuals carefully proofread a print piece for mistakes before it is actually produced and distributed.

Broadcast Channels

Traditional broadcast communication channels used by foodservice operators include radio and television. Both are one-way communication tools. An advantage to broadcast advertisements is that they can add sound (radio) or sound and movement (television) to ads which cannot be achieved with traditional print advertising.

Foodservice operators should also recognize that, unlike print ads, both radio and television advertising is subject to the rules enforced by the Federal Trade and Communication (FTC); a topic addressed in detail later in this chapter.

Radio   Radio ads are relatively inexpensive to produce, and they can be extremely powerful. The radio station on which a radio ad is broadcast will help assist in the ad's production. But foodservice operators must make important decisions about their ad's placement and frequency as these factors will affect the cost of broadcasting the ad.

The three factors that most impact the broadcasting costs of radio ads are time of day, popularity of the radio station, and the number of broadcast **spots** purchased. Radio spots are typically 15 seconds, 30 seconds, 45 seconds, or 1 minute in length.

Key Term

Spot (broadcast): A multimedia advertisement that is broadcast at a specific time.

In addition to the length of time an ad is broadcast, the time of day at which a radio ad is aired is a significant factor in the ad's cost. Operators will pay a higher price per spot when the number of a radio station's listeners is highest. This is typically during the early morning (drive to work) hours and late afternoon (drive home from work) hours.

Radio stations very carefully monitor their number of listeners. Radio stations that typically draw a larger number of listeners will charge more for a spot than those radio stations that draw smaller numbers. Popular radio stations can consistently charge more for their spots than can those stations that are less popular.

In nearly all cases, radio stations will charge a lower per-spot price when a foodservice operator purchases a larger number of spots. Thus, the per spot cost for an operator who elected to run a radio ad 100 times will be lower than the per spot price paid by an operator who runs a similar-length ad only 10 times.

Radio ads can be of several types:

1) **Straight read:** In this type of ad, the radio station produces and records the ad. It utilizes one or more voice actors, provides the music and sound effects, and then broadcasts the ad during the times of the customer's purchased spots.
2) **Live read:** In this type of ad, the host of a live radio show reads the advertisement in real time during the show.
3) **Sponsorship:** In this type of ad, a business's name is mentioned as a sponsor of the radio program being broadcast. In this arrangement, there may also be a short explanation of the company sponsor. For example, the sponsorship ad for a food operation serving Italian foods may be: "This segment sponsored by Giuseppe's Italian Patio, home of the Endless Pasta Bowl."

Television   Unlike radio ads, broadcast television ads display visual messages in addition to audio features, and this can make televised ads extremely powerful and effective. Despite the ever-increasing growth of the internet and the changing viewing habits of customers, large numbers of people still regularly watch broadcast and cable television.

While television advertising is effective, it is also expensive. A professionally developed television advertisement will require the employment of one or more actors, a director, camera persons, a scriptwriter, and a video editor. Like radio spots, television spots are typically produced as 15-second, 30-second, 45-second, or 1-minute ads.

Also, like radio spots, the cost of television spots will vary based on the time of day they are broadcast. Television spots cost the most during the times most people are watching television (prime time).

Large foodservice chains typically produce ads that are shown on national television. Smaller operators most often produce ads that are shown on local or cable networks, or during local newscasts broadcast on national networks.

Regardless of whether they are producing radio or television ads, successful foodservice operators must remember the **AIDA** advertising tool.

Useful as a guide when creating traditional broadcast messages and other advertisements, AIDA stands for:

Key Term

AIDA: An advertising tool used as a guide to help ensure an advertisement is effective. AIDA is an acronym for attention, interest, desire, and action.

Attention (or attraction)
Interest
Desire
Action

The attention component of AIDA reminds marketing professionals that, in a media-filled world, an operator's radio or television ad must be quick and direct to grab the potential guests' attention. This involves using powerful words or pictures that catch the listener or viewer's attention and makes them stop and concentrate on what will be coming next in the ad.

The interest component of AIDA addresses the fact that consumers will only continue to pay attention if they understand that the ad is directed toward them. Therefore, the foodservice operator creating the ad must show that the products and services they will provide are the ones the listeners or viewers are interested in.

The goal of the AIDA's desire component is for the listener or viewer to think that "I want (or need!) what this advertiser is selling!" Recall that foodservice customers most often want to buy more than food and beverages. What they may also desire is convenience, an experience, an adventure, or to be treated in a very special way. The desire component of AIDA addresses these wants.

The action component of AIDA requires an advertiser to make specific recommendations about what listeners or viewers should do next. For foodservice operators, that may mean the advertisement is concluded with phrases such as:

> *Call now to make a reservation!*
> *For more information visit our website.*
> *Stop in today!*

A specific call to action can often prompt potential guests to follow the action-related suggestions offered in the broadcast ad.

Web-Based Communication Channels

All foodservice operators must carefully evaluate the myriad of traditional communication channels that can be used to sell food and beverage products. Increasingly, many consumers prefer to obtain their buying information from the Internet. As a result, all operators must carefully monitor currently popular web-based communication channels and those that are emerging in popularity.

A foodservice operation's most important web-based communication channel is its own proprietary website. The management of an operation's this website is so important that it will be the only topic in Chapter 9.

The owners and operators of a restaurant are not the only ones who use websites to communicate information about their operations. For example, the monitoring and management of information posted on third-party-operated websites and apps is also essential. The management of these types of sites is the sole topic of Chapter 10.

Finally, foodservice consumers increasingly look to websites in which consumers themselves post information. These **user-generated content (UGC) sites** must also be carefully managed, and strategies to do so are the sole topic of Chapter 11.

All types of web-based channels are typically interactive, and they rely on the internet or wireless communications to deliver their content. Web-based channels consist of several subcategories including websites, social media sites, and cell phone applications. Regardless of the web-based communication channel, operators can evaluate the effectiveness of these channels when **digital marketing metrics** are used.

There are several digital marketing metrics that are normally important to foodservice operators. Among the most important of these metrics are:

✓ Traffic
✓ Conversions
✓ Engagement

Traffic

Traffic on a digital site simply refers to the number of people who have visited the site. In most cases, the company hosting the site can provide an operator with regular reports of the number of visitors to their site on a daily, weekly, or monthly basis.

Key Term

User-generated content (UGC) site: A website in which content including images, videos, text, and audio have been posted online by the site's visitors.

Examples of currently popular UGCs include Instagram, Twitter, Facebook, Pinterest, and YouTube.

Key Term

Digital marketing metric: A key performance indicator (KPI) used to measure the success of a business's online marketing efforts. The goal of using digital marketing metrics is to track, record, and assess how consumers interact with a business online through websites and social media platforms.

For many foodservice operators who maintain a brick-and-mortar storefront, increased traffic on a digital communication channel should directly lead to increased in-store customer counts. As a result, operators can count the number of customers who entered their businesses before, during, and after a digital ad campaign. One good way to gain an understanding of a digital ad's effectiveness is to ask arriving customers how they heard about the business.

Another effective way to determine whether an increase in business is the direct result of a particular online ad campaign is to use a unique **promotional code** in the ad. For example, when a foodservice operator allows guests a price reduction of 10% by entering the word "SAVE" into a field in a digital ordering systems' checkout section, the word "SAVE" is an example of a promotional code. Customers could be asked to use this code when placing orders online or when they arrive on site. In either situation, the use of a promotional code is one good way to evaluate a site's traffic level.

Key Term

Promotional code: A series of letters or numbers that allow a consumer to receive a discount on a specific purchase.

Conversion

Conversions refer to responding to several distinct types of calls to action such as making a purchase, placing an order, or creating a personal account. While "traffic" refers to the number of visitors who see an ad, conversion measures the number of people who make a purchase or who follow some direction given in the ad.

It is easy to see that an ad may be viewed by many potential guests but could result in very few actual orders being placed. Alternatively, an ad may have lower traffic counts, but higher conversion rates. The conversion rate of an ad is one good indicator of the ad's effectiveness with customers.

Engagement

Engagement refers to the amount of time a user spends viewing an ad on a specific web-based communication channel. In most cases, the more time a viewer dwells on a website, the more effective the communication is. It is reasonable to assume that dwell time related to a specific web-based ad provides a reasonable estimate of the time users read or watch an ad. In most cases, the longer someone dwells on a web page the more likely they are to engage with it, develop an affinity for it, and return to the site.

Technology at Work

As is true with all advertising efforts, it is important that foodservice operators understand the effectiveness of their web-based advertising efforts.

In most cases, however, smaller foodservice operators will not have the technical skills required to carefully assess the KPIs of their digital marketing efforts. Fortunately, a large number of companies provide the services of monitoring and reporting on web and other digital platform marketing activities.

To review the types of services offered by specialists in digital marketing, enter "top restaurant digital marketing agencies" in your favorite search engine and view the results.

Assessing New Communication Channels

Experienced foodservice operators know that the number of new communication channels appearing on the market continues to increase steadily. These new channels may be traditional or web-based. In all cases, operators can help evaluate the wisdom of adopting these emerging channels by addressing some key questions.

1) Will the new channel likely be embraced by my own target market?

A new communication channel may be increasingly popular, but not of specific interest to an operation's target market. While broad generalizations should be avoided, it is true that some demographic segments of the population are more prone to use a new channel than are others.

For example, recent studies have shown that younger users are more likely to report using their smartphones and multiple smartphone apps on a daily basis than are older users. When older consumers use their smartphones, it is most often for e-mail, directions, online searches, or checking social media sites.

Similarly, a new radio station may feature classic oldies music from the 1960s and 1970s. While such a station may be very popular with today's senior citizens, that same radio station is less likely to be popular with listeners in the 18–25-year-old range. If an operation's target market are 18–25-year-old guests, choosing to place advertising messages on an "oldies" radio station would likely be a poor choice even if that station is very popular with its own target market.

2) Is there a method by which I can track the effectiveness of the new channel?

Tracking the effectiveness of a marketing channel is always a challenge for operators. For example, while a newspaper may be able to tell an operator how many copies are printed each day, it is not able to tell an operator how many people would read the operator's ad if it were placed in the newspaper.

In the case of web-based advertising options, operators must consider whether there are readily available and accurate digital marketing metrics available to help evaluate an ad's effectiveness.

Operators should understand that an effective ad should do more than immediately increase the number of customers coming to a foodservice operation. The best communication pieces encourage positive memories and good feelings that influence guests' behavior over time, and that will encourage customers to visit an operation at some time in the future.

3) What is the goal of the new channel?

Some foodservice operators believe that the goal of all advertising should be to increase customer counts. While an increase in customer counts is almost always desirable, in most cases, using a communication method to seek new customers will be 7–10 times more expensive than addressing a marketing message toward existing customers. Building sales through increased guest frequency, increased check average, and increased party size are most often the better ways for a foodservice operation to increase its revenue. Carefully assessing a new communication channel's ability to reach new guests versus speaking to current guests is an important consideration.

4) Can the new channel be adopted without a significant reduction in the amount of the marketing budget assigned to existing and effective channels?

Most foodservice operations budget a certain percentage of total revenue for marketing purposes. In general, industry experts suggest that a typical foodservice operation should allocate 3–6% of its total sales to its marketing budget. As a result, to stay within their budgets operators considering the adoption of a new channel must recognize that, perhaps, an adoption should result in a decrease in the amount spent in other communication channels.

Redirecting advertising dollars away from less effective to more effective communication channels is always a good idea. However, redirecting advertising dollars from very effective communication channels to a new channel whose effectiveness is unproven may be a poor idea.

What Would You Do? 4.1

"But we have placed an ad in the 'Welcome to Our Town' magazine for the last five years," said Michelle Austin, manager of Cluck in a Bucket, a quick service restaurant featuring fried chicken, and traditional sides such as mashed potatoes and gravy, coleslaw, and biscuits.

Michelle was talking to David Berger, the operation's assistant manager. David had proposed using the advertising budget previously spent for an ad in the local Visitors' Bureau magazine to update the restaurant's website. While Cluck in a Bucket's website was professionally done, it did not show any videos of the operation's products, and it contained only print messages and photos.

"I think it's more important to keep our website up-to-date than to keep advertising in a print magazine. I know we should support the local visitors' bureau's efforts to promote our town, but I think it's more important to promote ourselves!" said David.

Assume you were Michelle. How would you go about deciding whether to continue placing a print ad in a community-wide magazine versus investing the same amount of money in updating your website? How important would it be for Michelle to carefully consider the implications of each alternative before making her decision?

Understanding Buyer Behavior

As they make decisions about the best communication channels for delivering marketing messages, foodservice operators must understand buyer behavior. First, they must recognize that there are four, rather than only one, types of buyers for their products and services. Then, they must understand the process all buyers go through as they make their purchase decisions.

The Four Buyer Types

In his book, *Free to Choose*, Noble Prize-winning Economist Milton Friedman suggested that to best understand customers, all businesses should recognize that they serve not one, but rather four very different types of buyers.[1] Each of these four types of buyers makes his own unique value judgments when considering a foodservice purchase.

---

1 https://www.allencheng.com/free-to-choose-book-summary-milton-friedman-rose-d-friedman/

Recall in Chapter 1 that we defined "value" for a foodservice guest as the amount paid for a product compared to the buyer's view of what was received for the price that was paid. This key relationship between the price paid and the value received can be illustrated by the following formula:

Price Paid – Value Received = Personal Profit

Each of the four buyer types suggested by Friedman has a personal profit formula as shown in Figure 4.3. Note that the personal profit formulas utilized by a buyer depend on two key factors: whose money is spent and who the money is spent on.

#1 Type Buyers

In the #1 type of buyers category, buyers make a personal judgment about spending their own money on themselves. For example, an office worker is given a certain amount of time off work to, at their expense, buy and consume lunch and then return to their office. In this example, the buyer is spending their own money on themselves.

#1 type buyers typically are very careful about how they spend their own money. These buyers seek to optimize the value received for what was spent because the benefit or personal profit goes directly to themselves. These are the buyers who want to get a good deal on whatever they are buying and are very proud of themselves when they are successful.

These buyers are often willing to compromise, for example, on designated times in which specially priced meals can be purchased in a restaurant. They are usually quite flexible in their buying behavior and are a prime market for strategic pricing decisions related to off-season or off-hour buying.

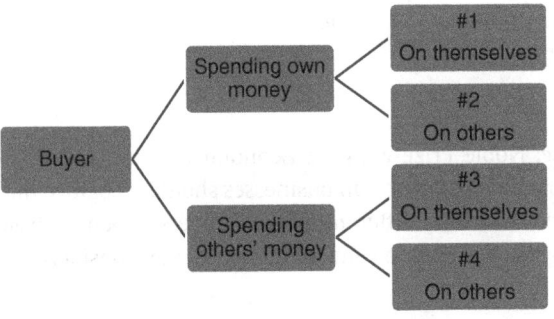

Figure 4.3 Four Types of Foodservice Buyers

#2 Type Buyers

#2 type buyers are spending their own money on someone else. The purchase may be made for a spouse, child, parent, significant other, or friend. While value is still of significant importance to these buyers, so too are perceptions of quality and even pride. These buyers want to receive good value, but they are typically less concerned about that than they are about the perceptions of the persons on whom they are spending their money.

For example, if an adult child is taking his or her mother out for lunch on Mother's Day, that foodservice buyer will be concerned about the costs. However, there is likely to be *more* concern about the mother's perception of the operation chosen and the product and service quality that was delivered.

No consumer wants to spend their own money to benefit someone else and then find that the money was poorly spent. #1 type buyers might boast about paying the lowest price possible for their purchase, but gift-givers (#2 type buyers) do not want their recipients to believe they were given the cheapest meal or lowest-cost service experience possible.

#3 Type Buyers

#3 type buyers are spending someone else's money on themselves. In a foodservice operation, this is the case when customers are reimbursed for their meals by their employer or by another entity.

These guests may have a reasonable restriction placed on the total amount they can spend. However, within those parameters of what they can spend they will be much less price-conscious than buyers spending their own money. Fine-dining and upscale foodservice operators understand this well and realize that many of their guests are likely to be #3 type buyers.

#3 type buyers are also very likely to use prepaid gift cards to restaurants that they have received from others for their birthdays, holidays, or other special event. Holders of these prepaid gift cards consistently spend more money on average than those who buy their meals without such cards.

Find Out More

Gift cards continue to increase in popularity as increasing numbers of gift givers choose to use them.

Studies show that gift card recipients like receiving the cards, Additionally, when they use them, they tend to spend *more* than the value of the gift card. As a result, most foodservice operators will benefit if they offer both physical and digital gift cards as a choice to their customers.

To learn more about the buying behavior of gift card recipients enter "how consumers spend their gift cards" in your favorite browser and view the results.

When #3 type buyers can purchase appetizers, drinks, and desserts without undue concern for the price they frequently will do so. This is one reason corporate clientele on expense accounts are desirable clients for many foodservice operators.

#4 Type Buyers

#4 type buyers are spending someone else's money on someone else. Like #3 type buyers, these buyers bear little or no personal responsibility for the purchases they make. As a result, they are likely to be less price-conscious than buyer types #1 and #2. This is the case, for example, when representatives from a company or organization contact a foodservice operator to plan the company's holiday party or other special event.

Meeting planners and wedding planners are additional examples of #4 type buyers. These professional buyers may often be extremely cautious when spending money. One reason is that they know their purchasing choices will be evaluated based on the value perceptions of *others*, rather than merely their own.

Key Term

Meeting planner: An individual responsible for securing the meeting space, food, and other services required by those attending group meetings.

Also known as a Meetings Professional or Event Planner.

Find Out More

Meeting Professionals International is the world's largest association of professional meeting and event planners.

Their members routinely choose restaurants to host their client's meetings, or to simply provide foodservice for meeting attendees. Meeting and event planners can be a tremendous source of business for many foodservice operations.

To better understand the profession of meeting and event planning and how meeting planners select the foodservice operations they want to do business with, enter "meeting planners international association" in your favorite browser and view the results.

Buyer Behavior

When buyer-customers make decisions regarding the foodservice operations they will choose, an identifiable decision-making process takes place. This is true regardless of whether a buyer is of the #1, #2, #3, or #4 type.

This decision-making process is affected by factors including the buyer's socio-economic level, friends, and family status. It is also impacted by the individual buyer's own needs, past experiences, personality, and attitude.

For buyers of a foodservice operation's products and services, the decision-making process used will typically involve five key steps:

Step 1: Need Recognition
Step 2: Information Gathering
Step 3: Information Assessment
Step 4: Purchase Decision
Step 5: Evaluation of Purchase Decision

Step 1: Need Recognition

Recall from Chapter 2 that a customer's need can be defined simply as something they lack, or the difference between where the customer is and where they would like to be. Recognition occurs when a buyer acknowledges that difference. For example, if a potential foodservice customer feels hungry while driving home from work, this may trigger a need to search for a restaurant to satisfy the driver's hunger.

Similarly, for a couple seeking to celebrate their anniversary, the perceived need may be to feel important and to be treated in a special way in a special environment. This may lead the couple to seek a unique or upscale restaurant to satisfy those needs.

Foodservice operators must recognize the wide variety of needs that their target market guests may want to satisfy when they dine out.

Step 2: Information Gathering

After a buyer recognizes a specific need, the buyer will next seek information about ways to meet that need. The search for information may mean recalling their own personal experiences, asking friends or relatives, or reviewing various media sources (either traditional and/or web-based).

A buyer's search for information may be brief or extensive. A hungry buyer may simply see an exterior sign for a foodservice operation and stop in. A bride-to-be planning a wedding reception at an upscale restaurant may take weeks, or even months to accumulate information before she makes a decision.

The most important thing for foodservice operators to understand about the information-gathering phase is that most consumers place *more* emphasis on personal experience and the thoughts of their friends and relatives than they do on advertisements placed by a business. This is one of the reasons for the ever-increasing use of social media review sites such as Facebook, Yelp, TripAdvisor, and Zagat as major sources of buyer information.

Step 3: Information Assessment

In most cases, a foodservice customer will have a fairly substantial number of alternative choices available as they attempt to meet their purchase needs. Each of these alternatives may have positive and negative attributes.

For example, one foodservice operation may be fairly inexpensive, and this is considered to be a positive characteristic by the buyer. However, that same operation may not be in a location that is convenient for the buyer, and this is a negative consideration.

In most cases, potential buyers will consider both product and service levels as they make their dining-out choices. It is for this reason that it is so important for foodservice operators to carefully choose the product or service characteristics they wish to emphasize in their brand identity messages (see Chapter 3).

Step 4: Purchase Decision

This is the step at which the buyer actually makes a buying decision. The decision will be based on the buyer's perceived risks associated with each alternative that is available to them.

For example, if a foodservice customer elects to visit a foodservice operation associated with a national chain that is well known and well thought of, the buyer's risk is relatively low. If, however, the same buyer chooses to make purchases at a foodservice operation with which they are not familiar, and about which they know very little, their risk will be perceived as relatively high. This is a reason why new and independent operations must work diligently to establish their brand identity in their buyers' minds. As they do so, they reduce some of the risk potential customers may feel about patronizing an operation where product quality and service quality is not well known.

Step 5: Evaluation of Purchase Decision

After making a purchase decision, a buyer's evaluation seeks to answer the question "Did the purchase I made satisfy my need and meet my expectations?"

In most cases, buyers of foodservice products will assess whether the products' quality met their expectations and whether service levels were as they anticipated. It is easy to see that if a foodservice operation does not deliver on the promises made in its marketing messages, their guests are likely to be disappointed.

For example, consider a pizza operator who claims in marketing messages that its pizzas will be ready in 30 minutes. However, the new guest discovers that it takes nearly an hour for their pizza to be prepared. That operation has failed to deliver on its promises, and the guest will be rightfully disappointed.

Experienced foodservice operators know that it is almost always best to under-promise and over-deliver. When that is done, the chances of meeting guest expectations increase, as does the likely willingness of guests to return in the future.

Operator Responsibilities in Target Market Messaging

Regardless of their marketing messages or the marketing channels used to deliver them, all foodservice operators have legal, ethical, and social responsibilities as they advertise their products and services. As hospitality professionals, understanding and addressing all of these challenges is essential.

Legal Responsibilities

While foodservice operators have a great deal of freedom in how they market their products and services, those freedoms are not absolute. In the United States, local, state, and federal governments all have enacted and will strictly enforce laws and regulations related to the marketing of food and beverage products.

For example, at the local level towns, cities, and counties may have specific rules about how and when alcoholic beverages may be served. This is so because, in 1933, The U.S. Congress repealed a previously passed Federal law (the Eighteenth Amendment to the U.S. Constitution). It prohibited the sale of alcohol and granted states, countries, and towns the ability to pass their own laws restricting, or if they choose, even prohibiting the sale of alcoholic beverages.

At the Federal level, the **Federal Trade Commission (FTC)** is the agency primarily responsible for ensuring businesses advertise in a way that is fair to consumers. The goal of the FTC is to protect consumers and competition by preventing anticompetitive, deceptive, and unfair business practices through law enforcement, advocacy, and education.

Key Term

Federal Trade Commission (FTC): The agency in the U.S. government responsible for protecting consumers and competition by preventing anticompetitive, deceptive, and unfair business practices.

Under FTC rules, all advertisements must be truthful, cannot be deceptive or unfair, and must be evidence-based. For example, if a foodservice operator claims that an entree contains 1000 calories, then there must be scientific evidence to back up that claim.

The FTC enforces these truth-in-advertising laws, and it applies the same standards no matter where an ad appears—in newspapers and magazines, online, in the mail, or on billboards or buses. Radio and television stations are required to ensure that commercials meet FTC standards before agreeing to broadcast them.

If the FTC finds a case of fraud perpetrated on consumers, the agency can file actions in federal court for immediate and permanent orders to stop the false ads to prevent the fraud from continuing. It can also freeze the assets of a business and even require compensation for victims of misleading advertising.

Key Term

Food and Drug Administration (FDA): The agency in the U.S. government responsible for assuring that foods (except for meat from livestock, poultry, and some egg products which are regulated by the USDA), are safe, wholesome, sanitary, and properly labeled.

Other federal agencies involved in regulating food and beverage advertising include the **Food and Drug Administration (FDA)**, which is responsible for matters related to food labeling, and the **Bureau of Alcohol, Tobacco, and Firearms (ATF)** which is responsible for matters related to alcohol labeling and advertising.

The **United States Department of Agriculture (USDA)** plays a very large role in monitoring what foodservice operators can and cannot say about the products they sell. For example, many of the terms used to describe menu items in a foodservice operation, such as "healthy," or "fat-free," are subject to USDA regulations. These regulations will be addressed in detail in Chapter 7.

Legislation related to the legal marketing of food and beverage products changes frequently, and it is essential that operators always remain current with laws that directly affect them in their marketing efforts.

Key Term

Bureau of Alcohol, Tobacco, and Firearms (ATF): The agency in the U.S. government responsible for, among other things, monitoring the manufacture and sale of alcoholic beverages.

Key Term

United States Department of Agriculture (USDA): The agency in the U.S. government responsible for regulating the production and sale of most food products manufactured and sold in the United States.

Ethical Responsibilities

Ethics is a very important topic for those responsible for marketing foodservice operations. Ethical behavior can be distinguished from legal behavior. Legal behavior is required by the laws of society. Ethical behavior is not dictated by law, but rather it is determined by what an individual or organization believes are the right, or proper, things to do when interacting with others.

Key Term

Ethics: The moral principles that govern a person or business organization's conduct toward others.

For those involved in marketing a foodservice organization, ethical behavior means doing the right thing for customers. While not every operator has the same definition of doing the "right thing" for customers, those responsible for an operation's marketing efforts can easily describe the "right thing" as the "fair" thing. Other terms related to proper marketing ethics include integrity, humility, and most importantly, honesty.

Ethics in marketing impacts more than merely the contents of advertisements and promotions. When foodservice operators exhibit ethical behavior in their marketing efforts, employees of the food operation notice that and use it to shape their own actions towards guests.

Ethics and marketing take on even greater importance when operators recognize that surveys show that over 80% of consumers trust their friends and family's advice over the advice of a business. Nearly 70% do not trust advertisements of any type, and 71% do not trust company-sponsored ads posted on social media networks.

This issue of lack of trust means that those marketing foodservice operations must understand that their readers and viewers will look at *all* ads with great skepticism. Therefore, it is essential that those marketing food operations ensure their advertising efforts will always be viewed as fair and honest. Most foodservice operators agree that trying to damage a competitor's image when advertising is unethical.

Marketing experts believe that any fraudulent claims to customers, stereotyping of guests, or specific targeting of vulnerable audiences such as the elderly or children is unethical. For example, ethical advertisers know that young children typically have no money of their own to spend. As a result, targeting children in their ads and especially for foods that are not particularly healthy for them, is actually an effort to influence children to "pester" their parents to visit a particular operation. The more ethical approach would be to advertise directly to the adult parents, explaining exactly how the operation meets their dining needs and the real needs of their children.

Social Responsibilities

While ethical responsibilities in marketing relate to proper behavior toward individuals, social responsibilities in marketing relate to proper behavior toward society as a whole.

The **corporate social responsibilities (CSR)** of foodservice operators are gaining increasing attention of guests, and especially those guests who are environmentally conscious.

Key Term

Corporate social responsibility (CSR): Those business actions designed to enhance society and the environment instead of contributing negatively to them.

Find Out More

Putting corporate social responsibility (CSR) in action is good for business, even if it means making a decision not to do business. This was the path taken by the McDonald's corporation in the spring of 2022 when it decided to pull all of its stores out of Russia in response to the Russian invasion of Ukraine.

After having been in business in Russia for more than 32 years, the McDonald's corporation determined that it could no longer justify doing business with a country that was taking an aggressive military stance against one of its neighbors. Industry experts at the time estimated that it would cost McDonald's approximately $1.4 billion to cover its exit costs.

To find out more about this significant CSR-related decision that illustrates the importance of doing the right thing about corporate social responsibilities, enter "McDonald's leaves Russia 2022" in your favorite browser and view the results.

CSR actions are those activities that demonstrate a foodservice operator's commitment to do the right things for employees, the community, and the natural environment. Supporting local ranchers and farmers is one way foodservice operators demonstrate their CSR commitment. Respecting diversity, contributing to local and national charitable causes, adopting sustainable practices, and encouraging healthy lifestyles for guests are among other ways that foodservice operations demonstrate their CSR.

Demonstrating its CSR activities is good for business. For example, when a foodservice operation advertises that it is committed to using eco-sustainable packaging, it is employing CSR as a brand identifier that will appeal to many of its environmentally conscious guests.

Find Out More

Foodservice guests are increasingly interested in the way a foodservice operation relates to the environment. A food operation's CSR practices are increasingly a method by which foodservice guests select the operations they will frequent. Many of these guests are extremely passionate about their commitment to giving their business only to those foodservice operations that actively demonstrate CSR activities.

To see some examples of consumers explaining why they want to emphasize a foodservice operation's CSR activities, go to YouTube, and enter "why I prefer sustainable restaurants" in the YouTube search bar and view the results.

There is an old foodservice adage that states: "If you fail to plan you must plan to fail." A detailed marketing plan should be created after a foodservice operation has made decisions defining its brand and after carefully reviewing alternative communication channels to ensure its marketing message reaches the target audience. This plan will specify what marketing activities are to be undertaken, who will be responsible for them, and when they will be completed. Creating and implementing this formal marketing plan is the topic of this book's next chapter.

What Would You Do? 4.2

"Let me see if I understand this," said Carla, the owner of the Cajun Lunch Box, a Po-Boy shop that featured roast beef, baked ham, fried shrimp, and fried catfish sandwiches dressed with sliced tomato, shredded lettuce, and pickles.

Carla's roast beef and baked ham po-boys were very popular with her customers and most of these sandwiches were ordered with melted cheese. Carla was talking to Resse Davis, a salesperson with American Foods, Carla's major food supplier.

"It's simple," interrupted Resse, "If you buy our private-labeled processed pizza cheese you can melt it on your sandwiches and your customers will never know any difference."

"So it's a processed cheese food, rather than the 100% cheese I use now," said Carla.

"That's right," said Reese, "and it costs 20% less than the cheese you currently buy. It's a good deal for you."

Assume you were Carla. What brand identity issues are at play in this situation? What legal and ethical considerations might be in play?

Key Terms

Web-based communication channel

Spot (broadcast)

AIDA

User-generated content (UGC) site

Digital marketing metric

Promotional code

Meeting planner

Federal Trade Commission (FTC)

Food and Drug Administration (FDA)

Bureau of Alcohol, Tobacco, and Firearms (ATF)

United States Department of Agriculture (USDA)

Ethics

Corporate social responsibility (CSR)

Operator's 10-Point Tactics for Success Checklist

Evaluate your need for, and the current status of, each of the following operational tactics. For those tactics you think are important, but not yet in place, develop an action plan for its implementation including who will be responsible for the tactic's completion and the target date by which it should be completed.

Tactic	Don't Agree (Not Done)	Agree (Done)	Agree (Not Done)	Who Is Responsible?	Target Completion Date
				If Not Done	
1) Operator has considered the importance of traditional print-based communication channels when communicating with their target markets.	——	——	——		
2) Operator has considered the importance of traditional broadcast-based communication channels when communicating with their target markets.	——	——	——		

(Continued)

				If Not Done	
Tactic	**Don't Agree (Not Done)**	**Agree (Done)**	**Agree (Not Done)**	**Who Is Responsible?**	**Target Completion Date**
3) Operator has considered the importance of web-based communication channels when communicating with their target markets.	____	____	____		
4) Operator has utilized a specific process for identifying those communication channels that will best fit their own operations as they deliver their marketing messages.	____	____	____		
5) Operator has a system in place for monitoring and assessing new and emerging communication channels.	____	____	____		
6) Operator has reviewed and understands the differences in motivation exhibited by the four primary types of buyers.	____	____	____		
7) Operator has considered the step-by-step process all buyers go through as the make their purchase decisions.	____	____	____		
8) Operator has carefully reviewed their legal responsibilities related to marketing their own operations.	____	____	____		
9) Operator has carefully reviewed their ethical responsibilities related to marketing their own operations.	____	____	____		
10) Operator has carefully reviewed their social responsibilities related to marketing their own operations.	____	____	____		

5

Creating the Marketing Plan

What You Will Learn

1) The Importance of a Marketing Plan
2) How to Create a Marketing Plan
3) How to Implement a Marketing Plan
4) How to Evaluate the Results of a Marketing Plan

Operator's Brief

In this chapter, you will learn that foodservice operators should plan their marketing efforts. The best way to do this is to develop a formal and written marketing plan. There are important advantages that result from developing a formal marketing plan. These include improved assessment of the business environment, more focused decision making, better financial planning, more specific accountability, and overall assessment of marketing efforts. While there are some challenges to producing marketing plans, none should prevent a foodservice operator from undertaking formal market planning.

Regardless of its size, all foodservice operations should develop a formal marketing plan. The creation of a marketing plan requires five basic steps. These are (i) determining marketing strategies or goals, (ii) identifying target markets, (iii) determining the marketing budget, (iv) selecting marketing channels, and (v) making marketing task assignments.

When creating their marketing plans, foodservice operators must first identify their marketing strategies or goals. Then specific tactics to achieve the goals can be identified, and cost estimates required to implement the tactics can be established.

To implement their marketing plans, foodservice operators identify what marketing tasks will be done, when they will be done, and who is responsible for doing them. In most cases, the effective implementation of a marketing plan requires the efforts of management and the entire foodservice operation team.

After a marketing plan has been implemented, foodservice operators must evaluate the effectiveness of their efforts. In some cases, the results of a marketing plan's implementation exceed management's expectations and, in other cases, marketing efforts may not achieve the expected results. Then operators can carefully evaluate their plans to see where improvements can be made and where modifications to previously planned activities should be undertaken.

CHAPTER OUTLINE

The Need for a Written Marketing Plan
    Advantages of Developing a Formal Marketing Plan
    Challenges in Marketing Plan Development
Creating the Marketing Plan
    Identification of Marketing Strategies
    Identification of Marketing Tactics
    Creation of Marketing Plan Cost Estimates
Implementing the Marketing Plan
    Determining What Will Be Done
    Determining When It Will Be Done
    Determining Who Will Do It
    Importance of Staff Efforts to Implement Marketing Plans
Evaluation of Marketing Plan Results

The Need for a Written Marketing Plan

The effective marketing of a foodservice operation requires a formal plan of action. A **marketing plan** is a document foodservice operators prepare to guide them in their marketing efforts. The purpose of a marketing plan is to identify the strategies and tactics a foodservice operation will use to achieve its marketing goals. In most foodservice operations, the marketing plan is prepared annually.

Key Term

Marketing plan: A written plan detailing the marketing efforts of a foodservice operation for a specific time period (usually annually).

Marketing strategies are the broad and long-term goals a foodservice operation wants to achieve. These strategies may also be called marketing goals or objectives. **Marketing tactics** are the specific steps and actions an operation takes to achieve its marketing strategies. As shown in Figure 5.1, each marketing strategy (goal) identified helps foodservice operators choose the various marketing tactics they will need to employ to achieve it.

Simply stated, a marketing strategy describes *what* a foodservice operation wants to achieve, and its marketing tactics detail *how* the strategy will be achieved. Marketing strategies are first analyzed and then marketing tactics to achieve the strategies are considered.

Key Term

Marketing strategies: The broad and long-term marketing goals a foodservice operation wishes to achieve.

Key Term

Marketing tactics: The specific steps and actions undertaken to achieve a marketing strategy (goal).

For example, one marketing strategy for a business may be to increase its take-out sales in the coming year by 5%. The various marketing tactics that will be used to achieve that sales increase include specific advertising and promotional efforts. These advertising and promotional efforts must be carefully selected and implemented to help the operation meet its revenue increase goal.

Marketing a foodservice operation in today's competitive environment is simply too complex to be attempted without a formal marketing plan. For some foodservice operations, a critical strategy will be to obtain new customers. For others, it may be to strengthen relationships with current customers or to increase brand awareness. For still others, it may be to implement a completely new menu.

In most cases, a marketing plan is developed to cover a calendar year, or a large portion of a year. Since the marketing needs of individual foodservice operations vary, however, no "one-size-fits-all" marketing plan can be implemented. Foodservice operators must develop, implement, and evaluate their own unique marketing plans.

Figure 5.1 Marketing Strategies/Goals Dictate Marketing Tactics

Advantages of Developing a Formal Marketing Plan

A well-conceived and executed marketing plan can yield a tremendously positive impact on a foodservice operation and its profitability. By definition, a marketing plan indicates future actions. While no foodservice operator can predict the future perfectly, there are several advantages that can result from developing a formal marketing plan. These include:

1) Improved assessment of the business environment

   The creation of a formal business plan requires foodservice operators to carefully consider the needs of their customers, the abilities of their competitors, and numerous other factors that directly impact their business. These include an assessment of the social, economic, legal, and technological issues that directly affect how their businesses will be operated.

   A careful examination of the overall business environment helps foodservice operators identify realistic and achievable marketing goals and strategies. A formal assessment of their target market needs (that may regularly change!) also helps ensure operators continually stay customer focused.

2) Improved decision making

   After foodservice operators carefully evaluate their business environments and the needs of their guests, they are in a better position to identify specific factors directly affecting their operations.

   For example, if guests are requesting items not currently on an operation's menu changes may be in order. If new items are to be added to the menu, how will this information be communicated to current and potential guests? As explained in the previous chapter, foodservice operators can choose from a wide variety of communication channels as they send their marketing messages. In this example, some alternatives may be better suited for communicating the introduction of new menu items than other channels.

   When marketing strategies have been carefully identified, it becomes easier to select the tactics that directly support those strategies. Given many alternative choices in their marketing efforts, operators who create marketing plans can better decide which specific tactics will contribute most to the achievement of their marketing goals.

3) Better financial planning

   Since formal marketing plans will identify the timing of specific marketing activities and their costs, foodservice operators who formalize marketing plans can better manage their budgets. For example, assume that a foodservice operator knows that the busiest time of the year is the summer, and their slowest time is the winter. Formal marketing plans can schedule high-cost advertising expenses in the summer months when revenues are strongest, and lower-cost marketing activities might be planned for the winter months.

4) Improved accountability

In addition to identifying marketing activities to be undertaken, a formal marketing plan identifies the specific individual or team responsible for implementing marketing activities. As a result, improvements in accountability for management or marketing staff performance are inherent in developing a formal marketing plan.

The best marketing plans indicate what is to be done and when related activities are implemented. When specific marketing activities include the timeframe for completion, the accomplishment of planned activities as scheduled aids in achieving an operation's marketing goals.

5) Improved assessment of marketing efforts

The strategic goals of a marketing plan should be realistic and measurable. When a strategic goal is realistic, a foodservice operation has a good chance of achieving it. Goals that are set unrealistically high can discourage operators because the goals simply cannot be attained regardless of their efforts. Strategic goals that are set too low, however, may cause operators to be complacent in their marketing efforts.

The strategic goals identified in a marketing plan must also be measurable. Goals such as "increase business" or "improve customer satisfaction" can only be meaningful when they are measurable. For example, in a formal marketing plan a goal of "increase business" would better be stated as "Increase business by 15% on Friday nights." That goal is measurable and an operator can determine whether marketing efforts helped achieve that goal.

Similarly, a goal of "improve customer satisfaction" might better be stated as "improve the operation's average star rating from three stars to four stars on Yelp, Facebook, and TripAdvisor by December 31st of this year."

Again, that goal is measurable, and an operator can readily determine whether the goal was achieved during the time period covered by the marketing plan. The best marketing plans make it easy for foodservice operators to monitor their actual performance against their budgeted or expected performance.

Challenges in Marketing Plan Development

Despite the many advantages that accrue from developing a marketing plan, some foodservice operators do not formally plan their marketing activities. Those operators who are hesitant to create a marketing plan typically cite one or more of the following objections:

1) Marketing plans take too much time to develop

Creating a formal marketing plan will take time. However, marketing plan development enables operators to consider decisions that are needed at some point in the future. For example, if an operator has decided to use the radio to communicate a marketing message, determining when the radio ad will be developed and played becomes an important decision to be made. Making these decisions in advance will typically take no longer than making them on the spur of the moment.

Experienced operators would agree that setting aside the time needed to develop a formal marketing plan is worth the time invested.

2) My operation is too small to need market planning

No foodservice operation is too small to benefit from effective market planning. In fact, a good argument could be made that the smaller the operation the more important formal market planning becomes. Large foodservice organizations and chain operations likely have one or more full-time staff who do nothing but work on marketing efforts. If a smaller independent operator is to successfully compete against these organizations, then effective market planning is essential.

Without an effective marketing plan, an operation's growth will likely slow. The operation may attract few new customers, and existing guests may not become aware of new products or services being offered. This, in turn, reduces revenues from frequent repeat customers.

3) My operation needs to stay flexible because the future is uncertain

This challenge is an argument *for* formal market planning. Since the future is uncertain, it is essential that foodservice operators have performance goals. These goals are stated in the strategies sections of the marketing plan. Recall that marketing tactics are then selected to support marketing strategies. While the tactics used to achieve an operation's marketing strategies and goals may vary, they must be identified. Then they can be modified as the business environment changes.

Of course, flexibility is essential in a world that is rapidly changing and in which new technologies and trends emerge almost daily. However, in most cases, an overemphasis on flexibility may mean an operation is pulled in too many different directions and can then become unsure of the specific steps necessary to attain goals.

A formal marketing plan helps foodservice operators better understand their guests, their competitors, and their overall business environment. Its goal is to help develop a marketing mix (see Chapter 1) that guests will find highly attractive and enticing. Despite the challenges involved, the development of a formal marketing plant is essential to the success of all foodservice operations.

Creating the Marketing Plan

While the marketing plan for a foodservice operation will be unique to that operation, the basic steps utilized to create the plan are the same:

1) Determining marketing goals
2) Identifying the target market(s)
3) Developing the marketing budget
4) Selecting marketing channels
5) Making task assignments

These steps are previewed in Figure 5.2.

Step 1: Determining marketing goals

A foodservice operation's marketing goals should be achievable and measurable. In the first step of marketing plan development, foodservice operators may develop a marketing plan that addresses a 6-month or 12-month period of time.

For many operations, an important marketing goal is to increase revenue. If an increase in revenue is determined to be a major goal, operators must determine how much additional revenue they seek to generate, the source of that revenue,

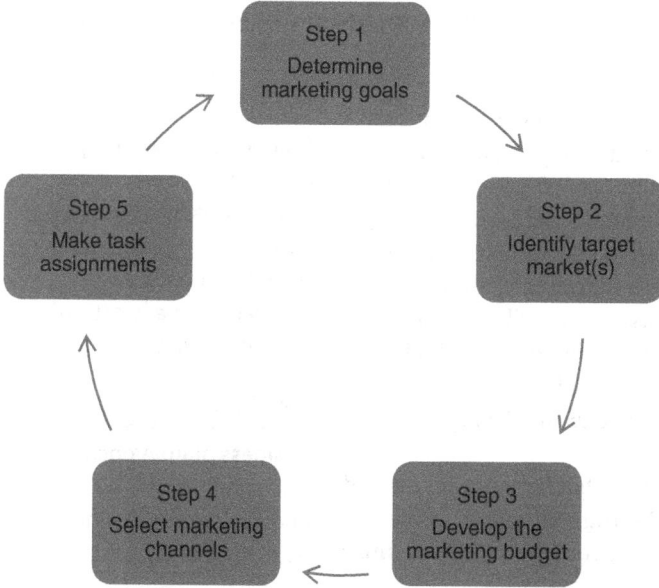

Figure 5.2 Steps in Marketing Plan Development

and the time frame by which it will be achieved. For example, a casual full-service restaurant may have a goal to "increase beverage sales by 10% when compared to the prior year." If management determines that this goal is reasonable, achievement of the goal can easily be measured.

The achievement of some marketing goals is more difficult to assess. For example, a foodservice operator may wish to "improve the ease of the online ordering system." While the achievement of this goal is more subjective, it can, however, be measured. One method is to use a consumer survey that requests users of online ordering systems to assess whether a newly implemented ordering process is easy to use.

Step 2: Identifying target markets

In this step, operators carefully review the target markets (see Chapter 2) they wish to serve. This is an especially crucial step for operators who identify more than one marketing goal. If this is the case, one form of the operation's marketing message may be directed to a specific target market, and another marketing message to another target market.

The identification of an operation's target market is essential because if substantial money and staff efforts focus on the wrong market, these financial resources will likely be misdirected and ineffective. Experienced foodservice operators know that their target markets can change. New target markets for foodservice operators may appear regularly, and their order of importance can change.

Step 3: Determining the marketing budget

In its simplest terms, a marketing budget reflects the amount or percentage of sales an operation wishes to devote to marketing efforts. For example, assume a foodservice operation forecasts sales of $1,000,000 during the time covered by the marketing plan and also assume it will spend 5% of its revenue on marketing. The marketing budget for the operation would then be $50,000 ($1,000,00 forecasted revenue × 0.05 marketing budget = $50,000).

For those operations that have not yet opened, **preopening marketing expense** forecasts will be identified in the operation's initial **business plan**.

Key Term

Preopening marketing expense: A nonrecurring promotional or advertising cost incurred before a new operation opens. Also referred to as a "start-up" marketing cost.

Key Term

Business plan: A document that summarizes the operational and financial objectives of a business.

A new foodservice business may require significantly more marketing effort and expense before it opens than would an existing and well-established operation.

While many foodservice advisors suggest that the average operation should spend 3–6% of its projected revenue on marketing, special circumstances may dictate that as much as 10%, or even more, of an operation's projected revenue might be devoted to marketing.

Step 4: Selecting marketing channels

In a previous chapter marketing channels were broadly categorized as being either traditional or web-based. In this step of marketing plan development, operators must select those specific marketing channels they feel are best able to deliver the marketing messages to be created.

Current popularity, cost, and ease of ad development are all factors that will impact an operator's decision about the specific marketing channels to select.

Step 5: Making task assignments

In very large foodservice operations, and in those with multiple units, one or more marketing specialists may be tasked with completing all of the operation's marketing efforts. In smaller operations, marketing tasks may fall to an operation's owner, manager, or assistant manager.

Task assignment is a critical aspect of marketing plan development because it identifies who will complete the marketing task to be undertaken, and when the task should be completed. It is in this stage of market plan development that operators will create their **marketing calendars**.

Key Term

Marketing calendar:
A schedule detailing all marketing activities planned for the time period covered by the marketing plan.

Marketing calendars should be specific, but they should also be flexible to allow an operator to change tactics as new circumstances dictate.

Find Out More

Every foodservice operator who develops a marketing plan should include a formal marketing calendar in the plan. Today's foodservice operators have many tools available as they create their marketing calendars. Whether they choose to produce their calendars using Google, Excel, or Word documents, templates that can help them get started can generally be downloaded from the Internet at low cost or no cost.

To see the many options available to foodservice operators, enter "free marketing calendar templates" in your favorite browser and review the results.

What Would You Do? 5.1

"I think we're too late," said Mateo.

"But it's only the middle of October," replied Gabriela. "Christmas is more than two months away. How can we be too late?"

Mateo and Gabriela were discussing potential holiday party bookings that would utilize their operation's banquet room. The banquet room could hold up to 150 people and was a perfect location for employee holiday parties hosted by businesses. One of Mateo and Gabriela's marketing goals for the year was to increase their number of holiday party bookings.

"Well," replied Mateo, "I talked to our marketing consultant today, and she said she could help us produce a very nice print ad for mailing to local corporations in only a week or two."

"So," replied Gabriela, "That's good. We can be sending the ad out by early November. What's the problem?"

"The problem is that the holiday meeting planners I've talked with have already booked their space for this year," said Mateo. "With most holiday parties being held as early as the first weekend in December, I think we should have started contacting our target market for holiday parties back in July or August at the latest!"

Assume you were Mateo and Gabriela. How could a formal marketing plan that included a calendar of marketing activities have helped to prevent the timing challenge they now face? What would you suggest they do for next year?

Identification of Marketing Strategies

Marketing strategies are the specific goals a foodservice operation wishes to achieve. Whether they are called strategies, goals, or objectives, they are the foundation of the marketing plan. The marketing goals of a foodservice operation can vary greatly based on the specific circumstances of its business.

While the specific marketing strategies selected by an operation will be unique to that operation, Figure 5.3 summarizes some commonly chosen foodservice operational strategies and presents their rationales.

It is important that foodservice operators understand that a strategy is not the same as a slogan or motto. For example, a stated strategy of becoming the "Best" coffee shop in a market area has little meaning and is subjective. Rather, a quantitative goal expressed in terms of customer counts or in monetary terms is much more useful. For example, "serving an average of 200 customers a day," or "generating $2,000 in sales per day" provide more useful and measurable goals.

Strategy	Rationale
Increase revenue	This strategy assumes that an operation can most easily grow its business by focusing on the marketing of the operation's currently existing products and services.
Increase profitability	This strategy assumes an operation can selectively promote those menu items and services that are most profitable, rather than focus on promoting those that are less profitable.
Increase brand awareness	This strategy is to build a business by expanding the number of potential customers who are familiar with the brand.
Increase **market share**	Regardless of market size, this strategy seeks to improve an operation's share of that defined market.
Generate new sales leads	This strategy is useful for operations that utilize personal selling to market their products and services. It expands the number of potential customers who become familiar with, and then buy, the operation's product and service offerings.
Successfully launch a new menu or menu item	This information-disseminating strategy is intended to increase a target market's awareness of an operation's new product offerings.
Enter a new market segment	This strategy seeks to communicate an operation's product and service offerings to the newly identified target market(s).
Increase website visitors	In most cases, as the number of visitors to an operation's website increases, its sales will also increase.
Increase activity on social media sites	Management of an operation's brand on social media sites is of increasing importance. This strategy utilizes social media to increase web users' interest in an operation and attract more guests.

Figure 5.3   Common Foodservice Operational Strategies and Rationale

Key Term

Market share: The percentage of the total category sales achieved by a single operation.

For example, if the sales of pizzas in a market area is 10,000 units per day, and one operation in the market sells 1,000 pizzas a day, that operation has a 10% market share (1,000 pizzas sold by the operation/10,000 pizzas sold by all operations = 0.10, or 10%).

When the goals are time-specific ("within six months, or "within one year"), a foodservice operator is better prepared to undertake the next part of market plan development; the identification of specific marketing tactics.

Identification of Marketing Tactics

For most foodservice operators, this is the part of market plan development that requires the most creativity and the greatest amount of insight into what customers want from the operation. It also requires the greatest knowledge of effective communication channels. The foodservice operator uses this section of the marketing plan to select the specific communication channels that will deliver the marketing message. (Recall from Chapter 3 that a marketing channel is simply the method, or means, of delivering a marketing message to an operation's target markets.)

For example, e-mail is a marketing channel. For many foodservice operators, e-mail is a simple and effective way of communicating with those customers from whom the operation has gathered e-mail addresses from those who have visited its website or connected in some other e-mail address gathering channel.

The use of radio is another effective communication channel. In both cases, foodservice operators seek to match the utilization of a specific communication channel to the specific strategy or goal they want to achieve.

To illustrate, if a goal of a foodservice operation is to launch a guest loyalty program, the utilization of e-mail would likely be a better marketing tactic than would be the use of radio. The reason: Customers who have provided their e-mails are already familiar with the operation. It follows logically that they are a better target market for a loyalty program than those who have not yet visited the operation. In this example, while radio could certainly be used to announce the formation of a loyalty program, it would likely be a less effective communication channel than the use of e-mails sent to known customers.

Technology at Work

While fewer and fewer foodservice operations find direct mail to be a cost-effective way of communicating with their customers, e-mail continues to be an excellent way to do so.

With the ever-increasing usage of smartphones and very easy access to e-mail inboxes, e-mail marketing should be a critical component of every operation's digital marketing strategy.

E-mails can be used to announce daily menus, introduce new menu items, or communicate special deals. They can also provide links to an operation's Facebook, Instagram, and Twitter accounts.

Since comprehensive e-mail marketing programs can be complex, many foodservice operators use an e-mail marketing service (EMS) for this purpose. An effective EMS can help an operation send notes, newsletters, and coupons to those on its e-mail list. It can also utilize a variety of techniques to help grow an operation's e-mail base.

To review some currently popular EMS companies and to see how they might assist in establishing a comprehensive e-mail program, enter "e-mail marketing services for restaurants" in your favorite search engine and view the results.

Marketing tactics are specific actions undertaken to support, or achieve, an identified marketing strategy. Figure 5.4 illustrates the identification of specific marketing tactics to be undertaken in support of a measurable marketing strategy.

In this example, a foodservice operation wants to increase revenue by offering discounts on appetizers during the traditionally slow period of 4:00 p.m. to 6:00 p.m. This is judged a useful tactic to attract more customers who may then stay for

Strategy/Goal	Tactics
Increase, within three months, total revenue achieved between 4:00 p.m. and 6:00 p.m. by 10% on a **year over year comparison** basis.	1) Produce a full-color table tent for in-house use, announcing discounted appetizers to be sold during the targeted time.
	2) Produce an e-mail announcing the appetizer promotion and send it out each week for three weeks to all those on the operation's current e-mail list.
	3) Create and post an ad on the operation's Facebook page announcing discounted appetizer prices and the times they are available.
	4) Create and post an ad on the operation's Instagram account announcing discounted appetizer prices and the times they are available.

Figure 5.4 Marketing Tactics to Support a Marketing Strategy/Goal

Key Term

Year-over-year (YOY) comparison: A method of evaluating two or more measured events to compare the results at one period with those of the comparable period of the previous year.

Also referred to as a "year-on-year" comparison.

dinner and/or purchase alcoholic beverages after the promotional period to help the operation achieve the increased revenue target.

As shown in Figure 5.4, the marketing tactics selected in a foodservice operation must be objective and specific, and they must be directly in support of an identified marketing goal. Each marketing strategy selected should have one or more specific marketing tactics identified in support of the selected strategy.

Creation of Marketing Plan Cost Estimates

In most cases, the completion of a marketing strategy included in a marketing plan requires an expenditure of funds. While an operation's marketing budget indicates the total amount to be spent on marketing efforts, marketing plan cost estimates detail the forecasted expense of implementing each individual marketing tactic.

To illustrate, assume that a foodservice operator determined that the annual marketing budget would be $95,000. Assume further that seven specific marketing tactics have been identified for the operation's annual marketing plan for the coming year.

As illustrated in Figure 5.5, the cost of implementing each of the seven chosen marketing tactics is identified, and the sum of those costs will be equal to the operation's overall marketing budget.

This step of market plan development may require operators to make difficult choices about the marketing tactics they will implement while staying within their overall marketing budgets.

No foodservice operation has an unlimited marketing budget. Therefore, operators often must make a distinction between marketing efforts that *could* be done versus marketing efforts that *should* be done within the resources allocated to marketing.

If an operation determines that its planned marketing activities exceed the amount budgeted for marketing, the operator has two choices. The first is to increase the total amount of the marketing budget, and the second

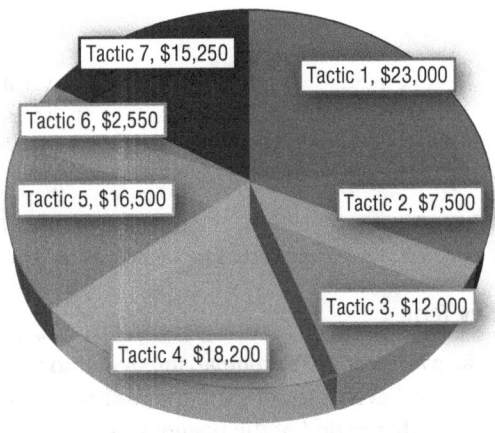

$95,000 Marketing budget

Tactic 7, $15,250
Tactic 1, $23,000
Tactic 6, $2,550
Tactic 5, $16,500
Tactic 2, $7,500
Tactic 3, $12,000
Tactic 4, $18,200

Figure 5.5 Marketing Budget Illustration

alternative is to decrease the amount of money spent on one or more of the other individual marketing tactics. These challenging decisions must be made, however, if the marketing plan is to be implemented within the operation's overall marketing cost limitations.

Implementing the Marketing Plan

As they finalize and implement marketing plans, foodservice operators address several key issues. Among the most important are determining:

✓ What will be done?
✓ When it will be done?
✓ Who will do it?

Determining What Will Be Done

When making the final determinations of which marketing tactics will be included in their marketing plans, operators must ensure they stay within their marketing budgets and optimize the use of their marketing dollars.

In most cases, a marketing plan includes the use of multiple marketing channels. For example, an operation may implement marketing tactics that require the use of traditional marketing channels such as broadcast or print and a variety of web-based communication channels including websites, social media, and blogs. In many cases, an operation will also implement various on-site marketing activities such as the use of in-house promotions, revisions of existing menus, and the training of service staff.

As they make final determinations about marketing channels to be used, foodservice operators must ask themselves several key questions about each potential channel:

1) Will it effectively reach my target audience?
2) Will it be cost-effective given the amount budgeted for its implementation?
3) Does it effectively communicate the marketing message?
4) Can its results be measured?
5) Is its use complementary to other marketing channels selected for use?

When making final decisions about the marketing tactics to be included in their marketing plans operators must continually recognize the difference between branding and marketing. While branding and marketing activities can overlap to some degree, recall from Chapter 3 that branding relates to specific ways a foodservice operation differentiates itself from its competitors.

Recall also that traditional and digital messaging can be important brand identifiers, but they are not the only brand identifiers. Other brand identifiers can include an operation's name, logo, exterior building design, interior design, décor, music, menus, employee uniforms, pricing structure, and services offered.

One good way to understand the relationship between branding and marketing is to recognize that branding is the way an operation differentiates itself from its competitors. In contrast, marketing includes the methods operators use to increase brand awareness among their target markets. As they finalize their marketing plans, foodservice operators must ensure that all marketing activities lead to increased brand awareness among their target markets.

Determining When It Will Be Done

As shown in Figure 5.4, each marketing strategy (goal) in a marketing plan will require that one or more specific tactics be completed to achieve the goal. Each chosen tactic should have a specific target completion date. This is important to monitor and ensure marketing activities are completed on time.

Figure 5.6 illustrates the use of a target completion date assigned to a specific marketing tactic chosen to support a specific strategic marketing goal.

Determining Who Will Do It

In some cases, the completion of a marketing tactic will require the efforts of a single individual. This would be the case, for example, when an operation's owner chooses to work alone with a radio station's salesperson to produce advertising spots that will be broadcast during the time covered by the marketing plan.

In other cases, however, the completion of a marketing tactic may require a team effort. This would be the case, for example, when a marketing tactic involves the introduction of a new menu item that will be promoted for a specific time. The successful introduction of the new menu item will likely require the efforts of the

Strategy/Goal	Tactics	Target Completion Date
Increase, within three months, total revenue generated between 4:00 p.m. and 6:00 p.m. by 10% on a year-over-year comparison basis.	1) Produce a full-color table tent for in-house use, announcing discounted appetizers to be sold during the targeted time period.	June 1, 20xx; one month prior to this promotion's implementation so staff can be properly trained in its implementation.

Figure 5.6 Marketing Tactics Completion Schedule

operation's purchasing staff (to ensure the item's ingredients are in inventory) and the kitchen production staff (to ensure proper training so the item is prepared properly). The service staff must also be able to address guests' questions about the new menu item's preparation style and/or its ingredients.

When the implementation of a selected marketing tactic involves the use of a team, it is best to designate a team leader who will be responsible for coordinating the efforts of those involved in the successful completion of the tactic. In all cases, the implementation of a selected marketing tactic should be implemented or coordinated by a single individual who can then be held accountable for the tactic's implementation.

Importance of Staff Efforts to Implement Marketing Plans

Each foodservice operation is different. It can be difficult to identify specific steps needed to ensure consistent staff support for the overall implementation of the marketing plan. In all cases, however, a successful marketing plan helps attract new customers and/or increases the frequency at which existing customers return. To attain these goals, each guest must receive a positive dining experience.

Experienced foodservice operators know that the staff who come in contact with guests are most responsible for managing the **moments of truth** their guests will experience. A moment of truth is every point in time when guests draw their own conclusions about the quality of a foodservice brand.

Key Term

Moment of truth: A point of interaction between a guest and a foodservice operation that enables the guest to form an impression about the business.

For example, if a foodservice operation's marketing efforts result in a potential guest visiting the operation's website to place an online order, then the website must be easy to use and functional. If it is not, guests will likely conclude that the operation is not service oriented. In this digital marketing example, the guest's moment of truth was most impacted by the actions of management as it developed the operation's website and not the operation's marketing plan.

If foodservice operations offer table service alternatives, the service staff personally manage moments of truth as guests interact directly with an operation's service staff. While each foodservice operation is unique, in many operations, staff efforts to support the marketing plan occur during a five-step process:

1) Greeting guests
2) Taking guests' orders
3) Delivering guests' orders
4) Serving guests as they enjoy meals
5) Collecting payment

The effective management of each of these key steps is critical if an operation's staff is to deliver on the promises made in their operations' marketing messages. As a result, each step is worthy of management's careful attention.

Greeting Guests

A friendly and professional initial greeting for arriving guests is important. However, foodservice operators should recognize that a guest's first moments of truth actually may occur well before they enter an operation's dining area or its carry-out pick-up station.

Guests will form an initial opinion of an operation when they visit the operation's website, arrive in the operation's parking lot, and as they approach the operation's main entrance. In each of these cases, guests will assess the quality of the operation even before they ever sit down in the dining room.

The physical appearance of a facility including clean, well-lighted parking and walk-up areas, is essential in forming positive guest perceptions about an operation's quality and the operation's attention to detail. The physical appearance of an operation gives guests a good idea about what they can expect to find inside. Moments of truth occur each time guests visually inspect the cleanliness of an operation's windows, its foyer and waiting areas, and even its guest restrooms.

Additional visual clues guests will encounter relate to the appearance of staff and their uniforms. Regardless of the level of formality in the operation, employee uniforms should be clean, fit properly, and include an easy-to-read name tag. Guests who observe employees in dirty, wrinkled, or inappropriate uniforms will draw negative conclusions about the quality of food and service they are about to receive. For that reason, an operation's policies on issues such as hair color and styles, visible tattoos, and body piercings must make sense and fit with the operation's overall brand image.

Whether they arrive at an upscale operation's entrance foyer, or at a drive-through window, each guest's initial greeting by an operation's staff should reflect a friendly and hospitable attitude. When that occurs, guests will form a positive opinion of the operation. If it does not occur, guests will notice, and the operation will have to work very hard to recover and overcome guests' initial and negative moments of truth.

Taking Guests' Orders

While the actual method used to take a guest order will vary based on the type of operation, it is important for operators to know that proper order-taking is a skillful art. Service reflects on the quality of an operation's server-training programs and management's commitment to its guests.

When staff have been trained to take orders politely, and to provide guests with accurate information regarding questions they may have about the menu, positive moments of truth will occur.

Inaccurate order-taking results in guests who do not receive the items they had planned to receive. When order taking is performed improperly, an operation's back-of-house kitchen staff may prepare ordered items perfectly, yet guests will still be disappointed in what they receive. Again, while an effective marketing plan may help attract guests to an operation, the service they receive will determine whether they have a positive experience and if they will return.

Delivering Guests' Orders

Today's foodservice guests can obtain their orders in a number of ways. The traditional method of delivering guest orders to those seated at tables is still prevalent in many foodservice operations. When guest orders are delivered this way, it is important that service staff get it right. Proper server training programs help ensure that all servers have a system in place to ensure that the right order is delivered to the right guest.

The proper delivery of guest orders is even more important when guest pick-up and drive-through delivery systems are in place. This is so because in these two delivery methods, guests will not actually see their individual menu items at the time they are received. Rather, the menu items will likely be packaged in ways that may make it difficult for guests to know if they have received the correct order until they return home or to another location. Then it is too late to correct any order errors, and guests will experience a negative moment of truth.

In increasing numbers of cases, the delivery of guest orders may involve the use of a third-party delivery partner (see Chapter 1) and these third-party relationships will be examined in detail in Chapter 10. However, operators must understand that moments of truth for a guest occur regardless of whether their orders are delivered by a third party or by the operation itself.

Serving Guests as They Enjoy Meals

Operators are aware that well-trained service staff can be a significant determinant when table service is used because service extends beyond providing the initial order. Examples may include the need to "check back" to confirm all is well with the guests' orders, to determine if a guest wants a beverage order (including water, soft drink, or coffee), and to suggestively sell a dessert.

The best servers do not normally ask, "Is everything okay?" Instead, a more friendly statement (example: "I hope everyone is enjoying their meal, is there anything else I can get you?") can inform guests that the server is available to provide additional assistance.

Experienced operators know the importance of a well-trained service team that works together. For example, when refreshing coffee or water, a server can assist more than just the customers in his/her own work section. If a problem is noted, the appropriate server can be informed about assistance that is needed. A manager can also circulate among customers and make new friends while providing additional beverages.

Collecting Payment

An operation's guest payment system has the potential to produce positive or negative moments of truth. For those operations providing on-site dining a negative moment of truth can occur when guests must wait excessively long to receive their checks, or when presented checks take an excessive time to be processed for payment by their servers. Alternatively, if guest checks are presented too quickly, on-site diners may feel they are being pressured to leave the restaurant. Well-trained servers can anticipate their guests' preferences and present their checks to them at the appropriate time.

Even when delivered at the proper time, a guest check that contains errors (for example when guests are charged for items not ordered or incorrect prices are charged for ordered items), negative moments of truth are the inevitable result.

It is easy to see that a foodservice operation could be excellent at marketing but may not be excellent at serving guests. Alternatively, an operation that provides an outstanding dining experience could be marketed poorly.

In the best foodservice operations, effective marketing plans draw guests to the operation, and excellence in food preparation and service produces positive moments of truth that will better ensure guests come back in the future. Therefore, it is essential that a foodservice operation's entire staff be made to understand their importance to the overall marketing effort.

Technology at Work

Effective marketing programs can attract new guests to a foodservice operation. However, the ways in which on-site staff manage their guests' critical moments of truth, are often the determining factor impacting guest return to the foodservice operation.

One effective way to help ensure that staff have the skills necessary to successfully serve guests is to identify and implement an effective guest service training program.

All foodservice operations, from quick service restaurants to fine dining establishments, must provide an excellent experience for their guests. This experience includes everything from taking guests' orders to delivering selected menu items appropriately. High-quality guest service training can improve customer skills for all employees who interact directly with guests. The good news is that online customer service training tools can be obtained and utilized very inexpensively, or even for free.

To review some currently popular free-to-use guest service training programs for foodservice operations, enter "free service training for restaurant staff" in your favorite search engine and review the results.

Evaluation of Marketing Plan Results

Chapter 12 explains how operators evaluate their overall marketing efforts. However, evaluation of the operator's marketing plan is a more specific management process that examines only the extent to which marketing strategies and goals identified in a marketing plan were achieved.

When marketing goals are measurable, a foodservice operator can calculate the **variance** between targeted financial results and actual results. Variances between revenue (income) and marketing costs (expenses) should also be calculated.

Key Term

Variance: The difference between an operation's actual and estimated income or expense.

A revenue variance identifies the (dollar amount) difference between what a foodservice operator estimated would be generated from a marketing strategy and what the strategy actually generated.

The formula used to calculate a revenue variance is:

$$\text{Actual revenue} - \text{estimated revenue} = \text{revenue variance}$$

For example, assume that a foodservice operator's marketing plan estimates that a specific marketing effort would generate an additional $50,000 in revenue within the next six months. At the end of the six months, the operator calculates that $40,000 in additional revenue was generated.

Using the revenue variance formula, this operator's variance would be calculated as:

$$\text{Actual revenue} - \text{estimated revenue} = \text{revenue variance}$$

or

$$\$40,000 \text{ actual revenue} - \$50,000 \text{ estimated revenue} = -\$10,000 \text{ revenue variance}$$

When analyzing their marketing results some foodservice operators prefer to calculate a **percentage variance**.

The formula used to calculate a percentage variance is:

Key Term

Percentage variance: The percentage difference between an operation's actual and estimated income or expense.

$$\frac{\text{Actual revenue} - \text{estimated revenue}}{\text{Estimated revenue}} = \text{percentage variance}$$

As noted earlier, percentage variance in revenue (or expense) is obtained by subtracting estimated revenue from actual revenue and then dividing the resulting number by estimated revenue.

When calculating a percentage variance using numbers from the previous example, the formula would be

$$\frac{\$40,000 \text{ actual revenue} - \$50,000 \text{ estimated revenue}}{\$50,000 \text{ estimated revenue}} = \text{percentage variance}$$

Or

$$\frac{-\$10,000 \text{ variance}}{\$50,000 \text{ estimated revenue}} = -0.20 \text{ (decimal form) or} -20\% \text{ (common form)}$$

In this example, the foodservice operation overestimated its revenue resulting from this marketing strategy by 20%.

Note that when using the variance percentage formula, the resulting decimal form percentage can be converted to the more frequently used common form by moving the decimal point two places to the right or by multiplying it by 100. Also, observe that the calculation of a revenue percentage variance can result in either a negative or positive percentage.

For each measurable marketing tactic included in the marketing plan, foodservice operators should calculate the variance in either number of dollars or percentage variance of the actual results when compared to forecasted results.

When a marketing tactic produces greater results than forecasted, an operator may want to consider utilizing that tactic again. If, however, a tactic did not yield the expected results, the operator may want to reconsider, or eliminate, the use of that tactic in the future.

With a well-thought-out and written marketing plan in place, foodservice operators can utilize their marketing budgets to increase their customer base and grow their revenues. However, marketing plans are not the only tools foodservice operators have available to them as they optimize their revenue and profitability. An additional, and extremely effective , marketing tool available to all foodservice operators is the individual prices charged for their products and services. The effective use of this powerful marketing tool is the topic of the next chapter.

What Would You Do? 5.2

"And the best part," said Steven, "is that it doesn't cost you anything because you have no marketing expense when you partner with us. We do all the work for you for free!"

Steven Bart, the sales representative for Coupon-online, was talking to Stella, the owner/manager of Stella's Street Tacos. Steven had just finished explaining his company's coupon program.

"Let me see if I understand you," said Stella. "We are going to sell a coupon good for $25.00 worth of purchases from Stella's, and we will sell it for $20.00."

"That's right," replied Stephen, "Then we keep $10.00 of that $20.00, and you get $10.00. It's an even partnership, and every month you'll get a check from us. And lots of new customers!"

Assume you were advising Stella. Do you agree that the Coupon-online program requires no marketing expense? Would you advise Stella to enter a partnership with Coupon-online? Explain your answer.

Key Terms

Marketing plan	Business plan	Moments of truth
Marketing strategies	Marketing calendar	Variance
Marketing tactics	Market share	Percentage variance
Preopening marketing expense	Year-over-year (YOY) comparison	

Operator's 10-Point Tactics for Success Checklist

Evaluate your need for, and the current status of, each of the following operational tactics. For those tactics you think are important, but not yet in place, develop an action plan for their implementation including who will be responsible for the tactic's completion and the target date by which it should be completed.

Tactic	Don't Agree (Not Done)	Agree (Done)	Agree (Not Done)	If Not Done Who Is Responsible?	Target Completion Date
1) Operator understands the importance of creating a written marketing plan.	___	___	___		
2) Operator has carefully reviewed the advantages of implementing a formal marketing plan.	___	___	___		
3) Operator has considered the challenges inherent in market plan development.	___	___	___		

(Continued)

Tactic	Don't Agree (Not Done)	Agree (Done)	Agree (Not Done)	If Not Done	
				Who Is Responsible?	Target Completion Date
4) Operator has identified specific measurable marketing strategies to be included in the formal marketing plan.	_____	_____	_____		
5) Operator has identified one or more specific marketing tactics to support each marketing strategy included in the marketing plan.	_____	_____	_____		
6) Operator has summarized and listed all marketing tasks required to implement the marketing plan.	_____	_____	_____		
7) Operator has created a formal marketing calendar detailing when specific marketing tasks will be completed.	_____	_____	_____		
8) Operator has made specific assignments of work to those chosen to complete marketing tasks.	_____	_____	_____		
9) Operator recognizes the importance of the entire foodservice operations team in achieving market plan results.	_____	_____	_____		
10) Operator has reserved time at the end of each market planning period to compare actual market plan results with the plan's expected results.	_____	_____	_____		

6

The Importance of Price in Marketing Efforts

What You Will Learn

1) The Importance of Pricing in the Marketing Effort
2) The Factors Affecting Foodservice Prices
3) The Methods Used to Price Menu Items
4) How to Evaluate Pricing Efforts in Foodservice Operations

Operator's Brief

In this chapter, you will learn that pricing is an important marketing tool for all foodservice operations, but prices are viewed very differently by sellers and buyers. For operators, costs and profits are a prime concern when establishing selling prices. For guests, the value received is of utmost importance. Therefore, from a marketing perspective, foodservice operators must be concerned that their prices communicate real value to their guests.

Numerous factors influence the prices foodservice operators charge for their menu items. Among the most important of these are economic conditions, local competition, level of service, type of guest, and product quality and costs. Additional factors important to pricing include portion size, delivery style, meal period, location, and bundling; a pricing strategy that combines multiple menu items which are then sold at a price lower than that of the bundled items purchased separately.

When establishing menu prices, some foodservice operators use a product cost-based approach. For these operators, a menu item's cost of production relative to its price is of most concern. When using this pricing approach, menu items with lower product cost ratios are thought to be more desirable to sell than those with higher product cost ratios. Other operators use a more profit-oriented approach to establish their menu prices. When using this

(Continued)

type of approach, those menu items that provide high profits-per-sale are considered more desirable to sell than those with lower profit levels.

Regardless of the pricing approach used, foodservice operators should regularly evaluate their menus to identify the items that are the most popular and most profitable. Then they can modify or even eliminate poor selling or unprofitable items and better promote those items that are most popular and most profitable.

CHAPTER OUTLINE

Pricing for Profits
    The Importance of Pricing
    The Operator's View of Price
    The Guest's View of Price
Factors Affecting Menu Pricing
    Economic Conditions
    Local Competition
    Level of Service
    Type of Guest
    Product Quality
    Portion Size
    Delivery Method
    Meal Period
    Location
    Bundling
Methods of Food and Beverage Pricing
    Cost-based Pricing
    Contribution Margin-based Pricing
Evaluation of Pricing Efforts
    Menu Engineering
    Menu Modifications

Pricing for Profits

Chapter 1 indicated that "Price," along with Product, Place, and Promotion, is one of the 4 Ps of Marketing.

It is essential that foodservice operators price their menu items to help ensure the long-term profitability of their businesses. Experienced foodservice operators know that if menu prices are set too low, an operation may be popular but not profitable. Alternatively, if menu prices are set too high, the popularity of a foodservice operation will likely suffer because very few guests will frequent it regularly. As a result, operators who seek to set the prices for their menu items in a way that yields reasonable profits must understand pricing very well.

The Importance of Pricing

Price plays a large marketing role in the interactions between a foodservice operation and its guests. To best understand the importance of price it is essential that operators first understand that the term *price*, as used in the foodservice industry, has two separate definitions.

Note that in both uses (noun or verb) the concept of an *exchange* between a buyer and a seller is present and important. The foodservice operator gives up (exchanges) a menu item for the selling price of the menu item. The foodservice guest gives up (exchanges) the amount of the item's selling price for the menu item.

Foodservice operators can typically charge any price they want to charge, but potential guests are free to accept or reject the operator's opinion that the prices charged are fair and will provide good value to them.

From the perspective of guests, price is easily the most noticeable, powerful, and understandable part of the entire 4 Ps of the Marketing mix. In fact, in some cases, a foodservice operation will be chosen, or not chosen, by a foodservice guest based primarily on the prices it charges.

Price is a powerful marketing tool. To best understand pricing in the foodservice industry, it is important to first recognize that price is viewed very differently from the perspective of an operator (the seller) and a guest (the buyer).

The Operator's View of Price

In most cases, operators are free to decide what products they sell, where they will sell them, and how they communicate their prices to potential guests. Foodservice operators are free to propose their prices, but they also must face the possibility that potential guests may *not* embrace their value propositions.

In the foodservice industry, price and value are related concepts. It is often stated that the value of any item is equal to what a buyer will pay for it. If this is true, when a sale is *not* made, the buyer either believed the item was not worth the asking price or a lower-cost alternative was available to the buyer that was considered worth its asking price.

From the perspective of an operator, a fair price should be an amount equal to the operator's incurred costs, plus a reasonable desired profit.

Key Term

Price (noun): A measure of the value given up (exchanged) by a buyer and a seller in a business transaction.

For example: "The price of the chicken sandwich combo meal is $9.95."

Price (verb): To establish the value to be given up (exchanged) by a buyer and a seller in a business transaction.

For example: "We need to price the chicken sandwich combo meal."

Stated mathematically, that concept would be:

Item cost + Desired profit = Selling price

When calculating their item costs, operators consider the costs of food, beverage, labor, and all other costs required to operate their businesses. Hospitality business owners and accountants can calculate these costs quite accurately.

The question of "What is a reasonable desired profit?" is more subjective. In the foodservice industry, an operation's **profit margin** is the percentage of each buyer's revenue dollar the operation retains as profit. The higher the profit margin, the more profitable the operation.

As addressed in Chapter 3, noncommercial foodservice operations rarely have profits as their primary reason to exist. In contrast, most commercial foodservice operations want to generate a reasonable profit margin and doing so is critical to staying in business and receiving a fair return for the risks involved in investing money in the business. For these operations, operating costs incurred and profit desired are the two primary factors of most importance when establishing menu prices.

Key Term

Profit margin: The amount by which revenue in a foodservice operation exceeds its operating and other costs.

The Guest's View of Price

In Chapter 1, "value" was defined as the amount paid for a product or service compared to a buyer's view of what they receive in return. Stated mathematically, that concept is:

Buyer's perceived benefit − Price = Value

When making a purchase, buyers want to receive *more* value than the value of what they are giving up. Stated another way, there are three possible buyer reactions to any seller's proposed selling price. These possible reactions are shown in Figure 6.1.

Buyer Assessment	Purchase Decision
1. Perceived benefit − price = a value less than "0"	Do not buy
2. Perceived benefit − price = a value equal to "0"	Do not buy in most cases
3. Perceived benefit − price = a value greater than "0"	Buy

Figure 6.1 Buyer's Assessment of a Seller's Value Proposition

When buyers believe the benefit they will receive is less than zero (less than what they give up in exchange), they generally will not buy. When the benefit is greater than zero, they are very likely to buy. When the perceived benefit is equal to 0, buyers are most often indifferent to the purchase. In this scenario, if no other alternatives are available, they may make a purchase decision. If alternatives are available, buyers will likely pursue those alternatives before making their purchase decisions.

To best understand how buyers' perceived benefit assessments directly impact their purchasing decisions, it is important to first understand the concept of **consumer rationality**.

Consumer rationality assumes that buyers consistently exhibit reasonable and purposeful behavior. That is, in most cases, buyers make purchase decisions based solely on their belief that it benefits them to do so.

Key Term

Consumer rationality: The tendency to make buying decisions based on the belief that the decisions are of personal benefit.

For foodservice operators, the acceptance of the concept of consumer rationality involves a willingness to look beyond the obvious and attempt to understand exactly how buyers believe they will benefit from a business transaction.

In some cases, this is not so easy. First, foodservice operators must resist the temptation to declare that their guests are irrational (for example, when operators criticize guests who state the operator's prices are "too high!")

All buyers like low prices, but what they seek most is value. Most foodservice guests are indifferent to the actual costs of operating a foodservice business and, therefore, they are also indifferent to an operator's profit margin. Also, it is important to recognize that rational buyers *do not* automatically equate a seller's price with the amount of value they receive in exchange. In fact, conventional wisdom advises them not to do so. From common sense and even from a legal perspective, buyers assessing a seller's value proposition are cautioned not to trust sellers. As a result, *Caveat Emptor*, the Latin phrase for "Let the buyer beware," is known and understood by most consumers.

Since many buyers may be skeptical about a seller's initial value proposition, it is the foodservice operator's responsibility when pricing menu items to ensure guests understand the answers to questions about "What do I get?" and "Why is it of value?" just as much as they understand the actual prices they will pay. Only then can buyers, who are increasingly very sophisticated and web-savvy consumers, be convinced they will consistently receive *more* than the worth of the money they must pay when making purchase decisions.

It is easy to see that a foodservice operator's primary motivation to recover operating costs and generate a profit is very different from their guests' motivation to optimize the value received for the prices they pay. Therefore, while operators must be concerned about their costs and profits, from a marketing perspective, they must be *more* concerned about utilizing price as a method of communicating excellent value to their guests.

Factors Affecting Menu Pricing

When foodservice operators find that profits are too low, they frequently question whether their prices (and thus their revenues) are too low. It is important to remember that the terms "revenue" and "price" are not the same thing.

"Revenue" means the amount of money spent by all guests and "price" refers to the amount charged for one menu item. Total revenue is generated by the following formula:

Price × Number sold = Total revenue

You can see that there are two components of total revenue. Price is one component, and the other is the number of items sold. Note: Generally (but not always), as an item's selling price increases, the number of that item sold decreases.

Experienced foodservice managers know that increasing prices without giving added value to guests does result in higher prices but, frequently, lower total revenue because of a reduction in the number of guest purchases. For this reason, menu prices must be evaluated based on their impact on the number of items that will be sold as well as their actual selling price.

While paying attention to costs and profits is always important, there are a number of other factors that directly affect the prices operators charge for the menu items they sell. Among the most important of these are:

✓ Economic conditions   ✓ Product quality   ✓ Location
✓ Local competition   ✓ Portion size   ✓ Bundling
✓ Level of service   ✓ Delivery style
✓ Type of guest   ✓ Meal period

Economic Conditions

The economic conditions that exist in a local area or even in an entire country can have a significant impact on the prices operators can charge for their menu items.

When a local economy is robust and growing, foodservice operators generally have a greater ability to charge higher prices for the items they sell. In contrast, when a local economy is in recession or is weakened by other events, an operator's ability to raise or even maintain prices in response to rising product costs may be more limited.

In most cases, foodservice operators will not have the ability to directly influence the strength of their local economies. It is their job, however, to monitor local economic conditions and to carefully consider these conditions when establishing menu prices.

Local Competition

While the prices charged by an operation's competitors can be important, this factor is often too closely monitored by the typical foodservice operator. It may seem to some operators that their average guest is concerned with low prices and nothing more. In reality, small variations in price generally make little difference in the buying behavior of the average guest.

For example, if a group of young professionals goes out for pizza and beer after work, the major determinant will not likely be whether the selling price for a beer is $6.95 in one operation or $7.95 in another. This small variation in price is simply not likely to be the major factor in determining which operation the group of young professionals will choose to visit. Other factors such as quality and location will be more important.

The selling prices of potential competitors are of concern when establishing selling price but experienced operators understand that a specific operation can always sell a lesser quality product for a lesser price. The prices competitors charge for their products can be useful information in helping an operator arrive at their own selling prices. It should not, however, be the only determining factor in pricing decisions.

The most successful foodservice operators spend their time focusing on building guest value in their own operations and not attempting to mimic the efforts of the competition. Even though many operators feel their customers only want low prices, remember that consumers often associate higher prices with higher quality products and, therefore, with a better price/value relationship.

Level of Service

The service levels an operation provides its guests directly affect the prices the operation can charge. Most guests expect to pay more for the same product when service levels are higher. For example, the can of soda sold from a vending machine is generally less expensive than a similar-sized soda served by a member of the service staff in a sit-down restaurant.

Service levels can impact pricing directly and, as the personal level of service increases in an operation, selling prices may also be increased. Personal service ranges from the delivery of products as in a delivered pizza, to the decision to quicken service by increasing the number of servers in a busy dining room. This tactic improves service quality by reducing the number of guests each server assists.

These examples should not imply that extra income from increased menu prices must be used only to pay for extra labor required to increase service levels. Guests are willing to pay more for increased service levels. However, higher prices must also cover extra labor costs and provide for extra profit as well. Many hospitality operators can survive and thrive over the years because of an uncompromising commitment to high levels of guest service, and they can charge menu prices reflecting enhanced service levels.

Type of Guest

All guests want good value for their money. However, some guests are less price sensitive than others, and the definition of what represents good value can vary by the clientele served. Consider the pricing decisions of convenience store customers across the United States. In these facilities, food products such as premade sandwiches, fruit, drinks, cookies, and the like are often sold at relatively high prices. The guests these stores cater to, however, most value speed and convenience, and they are willing to pay premium prices for their purchases.

In a like manner, guests at a fine-dining steakhouse restaurant are less likely to respond negatively to small variations in drink prices than are guests at a neighborhood tavern. A thorough understanding of exactly who the potential guests are and what they value most is critical to the ongoing success of foodservice operators as they establish menu prices.

Product Quality

Chapter 3 explained that some foodservice operators make product quality a major part of their key marketing messages. In many cases, a guest's quality perception of a menu item offered for sale can range from very low to very high. These perceptions are the direct result of how guests view an operation's menu offerings.

These perceptions are directly affected by the quality of a menu item, but they should never be shaped by the guest's view of a menu item's wholesomeness or safety. All foods must be wholesome and safe to eat. Guests' perceptions of quality will be based on numerous factors of which only one is actual raw ingredient quality. Visual presentation, stated or implied grade of ingredients, portion size, and service level are additional factors that impact a guest's view of overall product quality.

To illustrate, consider that, when most foodservice guests think of a "hamburger" they think of a range of products. A "hamburger" may include a rather small burger patty placed on a regular bun, wrapped in waxed paper, served in a sack, and delivered through a drive-thru window. If so, guests' expectations of this hamburger's proper selling price will likely be low.

If, however, the guests' thoughts turn to an 8-ounce "**Wagyu beef**-burger" presented with avocado slices and alfalfa sprouts on a fresh-baked, toasted, and whole-grain bun and served for lunch in a white-tablecloth restaurant, the purchase price expectations of the guests will be much higher.

Key Term

Wagyu Beef: Beef from a Japanese breed of cattle that is highly prized for its marbling and flavor. In the Japanese language, "Wa" means Japanese, and "gyu" means cow.

In many cases, a foodservice operator can choose from a variety of quality levels and delivery methods when developing product specifications, and as they plan their menus and establish their prices. The decisions they make will have a direct impact on menu pricing.

For example, if a bar operator selects a very inexpensive Bourbon to make whiskey drinks, they will likely charge less for whiskey drinks made than another operator who selects a better brand. However, guest perceptions of the value received from those buying the lesser-cost whiskey drinks will likely be lower than guests served a higher-quality product.

To be successful, foodservice operators should select the product quality levels that best represent their target markets' anticipated desires as well as their operations' own pricing and profit goals.

Portion Size

Portion size most often plays a large role in determining a menu item's price. Great chefs are fond of saying that people "Eat with their eyes first!" This relates to presenting food that is visually appealing and it also impacts portion size and pricing.

A pasta entrée that fills an 8-inch plate may well be lost on an 11-inch plate. However, guests receiving the entrée on an 8-inch plate will likely perceive higher levels of value than those receiving the same entrée on an 11-inch plate even though the portion size and cost in both cases are identical.

Portion size, then, is a function of both product quantity *and* presentation. It is no secret why successful cafeteria chains use smaller-than-average dishes to plate their food. For their guests, the image of price to value when dishes appear full comes across loud and clear.

In some foodservice operations, and particularly in "all-you-care-to-eat" operations, the previously mentioned principle again holds true. The proper dish size is just as critical as the proper size scoop or ladle when serving the food. Of course, in a traditional table service operation, an operator must carefully control portion sizes because the larger the portion size, the higher the product costs.

Many of today's health-conscious consumers prefer lighter food with more choices in fruits and vegetables. The portion sizes of these items can often be boosted at a fairly low increase in cost. At the same time, average beverage sizes are increasing as are the portion size of many side items such as French fries. If these items are lower-cost items, this can be good news for the operator. However, it is still important to consider the costs for larger portion sizes.

Every menu item to be priced should be analyzed to help determine if the quantity (portion size) being served is the "optimum" quantity. Operators would, of course, like to serve that amount, but no more. The effect of portion size on most menu prices is significant, and it is the job of an operation's back-of-house staff to establish and maintain control over desired portion sizes.

Delivery Method

Delivery method has become an increasingly important factor in establishing menu prices. There are essentially four ways foodservice operators deliver purchased menu items to their guests:

1) Dine-in service: Menu items are delivered to guests at their on-site tables or other seating areas.
2) Pick up/carryout: Guests receive their menu items from drive-through windows or pick up their items from designated on-site carry-out areas.
3) Operator direct delivery: Menu items are delivered to guests by an operation's own delivery employees.
4) Third-party delivery: Menu items are delivered to guests by one or more third-party delivery partners (see Chapter 1) selected by the operation.

The delivery style that has the greatest impact on menu prices is that of third-party delivery. From the perspective of guests, the decision to utilize a third-party delivery company such as Grubhub, DoorDash, or Uber Eats means the guest is placing important value on convenience, as well as on the menu items they choose.

When guests order directly from a third-party delivery company, they actually pay for:

✓ The selling prices of their selected menu items
✓ A service fee charged by the delivery company for providing the service
✓ A delivery fee for food delivered
✓ A gratuity; an optional tip for the delivery driver

The COVID-19 era saw explosive growth in the use of third-party delivery companies as many restaurants either closed indoor dining areas or severely restricted their inside seating capacities. However, the services of third-party delivery companies are not free to restaurants.

Depending upon the specific arrangement made between a foodservice operation and its third-party delivery partner(s), the operation will pay between 10 and 30% of a guest's total bill to the third-party delivery company. Thus, for example, if a foodservice operation has a customer who utilizes a third-party delivery app, and who purchases $100 worth of menu items from the operation, the operation will actually receive only $70.00–$90.00 from the third-party delivery company.

Astute readers will recognize that the foodservice operation pays a significant fee to satisfy their customers' desire for convenience. For that reason, many foodservice operators charge different (higher) menu prices when items are ordered from a third-party delivery app or avoid the use of third-party delivery services entirely.

Meal Period

In some cases, guests pay more for an item served in the evening than for that same item served during a lunch period. Sometimes this is the result of a smaller "luncheon" portion size, but in other cases, the portion size and service levels may be the same in the evening as earlier in the day. This is true, for example, in buffet restaurants that charge a different price for lunch than they charge for dinner. Perhaps operators expect those on lunch break to spend less time in the operation and will eat less. Alternatively, they may believe their guests simply spend less for lunch than for dinner.

Foodservice operators must exercise caution in this area, however. Guests should clearly understand why a menu item's price changes with the time of day. If this cannot be answered to the guest's satisfaction, it may not be wise to implement a time-sensitive pricing structure.

Location

Chapter 3 indicated that some foodservice operators make their locations a major part of their key marketing messages, and location can be a major factor in price determination. This is illustrated, for example, by food facilities operated in themed amusement parks, movie theaters, and sports arenas.

Foodservice operators in these locations charge premium prices because they have a monopoly on the food sold to visitors. The only all-night diner on an interstate highway exit is in much the same situation. Contrast that with an operator who is just 1 of 10 similar seafood restaurants on a **restaurant row** in a seaside resort town. In this case, it is unlikely that an operation

Key Term

Restaurant row: A street or region well-known for having multiple foodservice operations within close proximity.

will be able to charge prices significantly higher than its competitors based solely on its location.

One cannot discount the value of an excellent restaurant location, and location alone can influence price in some cases. Location does not, however, guarantee long-term success. Location can be an asset or a liability. If it is an asset, menu prices may reflect that fact. If location is a liability, menu prices may need to be lower to attract a sufficient clientele to ensure the operation achieves its total revenue and profit goals.

Bundling

Bundling refers to the practice of selecting specific menu items and pricing them as a group (bundle) so that the single menu price of all the items purchased together is lower than if the items in the group were purchased individually.

The most common example of bundling is the combination meals (combo-meals) offered by quick-service restaurants. In many cases, these bundled meals consist of a sandwich, French fries, and a drink. Bundled meals, are often promoted as "combo meals" or "value meals," and are typically identified by a number for ease of ordering.

Key Term

Bundling: A pricing strategy that combines multiple menu items into a grouping which is then sold at a price lower than that of the bundled items purchased separately.

Bundled menu offerings are carefully designed to encourage guests to buy all menu items included in the bundle, rather than to separately purchase only one or two of the items. Bundled meals are typically priced very competitively so that a strong value perception is established in the guest's mind.

Find Out More

You have now been introduced to numerous factors that influence the prices foodservice operators charge for the items they sell. In the future, Leadership in Energy and Environmental Design (LEED) certification achieved by an operation may well constitute another such factor.

The LEED rating system developed by the U.S. Green Building Council (USGBC) evaluates facilities on a variety of standards. The rating system considers sustainability, water use efficiency, energy usage, air quality, construction and materials, and innovation.

Increasingly, many consumers are willing to pay more to dine in LEED-certified operations. In addition, LEED-certified buildings are healthier for workers and for diners. The LEED certification creates benefits for foodservice operators, employees, and guests. It will likely continue to be of increasing importance to guests.

To learn more about LEED certification in the foodservice industry, enter "LEED-certified restaurant standards" in your favorite browser and view the results.

What Would You Do? 6.1

We have to lower our prices because there is nothing else we can do!" said Ralph, director of operations for the seven-unit Blinky's Sub Shops. Blinky's was known for its modestly priced, but very high-quality, sandwiches and soups.

Business and profits were good, but now Ralph and Rachel, who was Blinky's director of marketing, were discussing the new $6.99 "Foot Long Deal" sandwich promotion that had just been announced by their major competitor, an extremely large chain of subshops that operated thousands of units nationally and internationally.

"They just decided to lower their prices to appeal to value-conscious customers," said Ralph.

"But how can they do that and still make money?" asked Rachel.

"There's always a less expensive variety of ham and cheese on the market," replied Ralph. "They use lower-quality ingredients than we do. We charge $8.99 for our foot-long sub. That wasn't bad when they sold theirs at $7.99. Our customers know we are worth the extra dollar. Now that they are at $6.99 ... I don't know, but I think this is really going to hurt us. You are in charge of marketing," said Ralph, "What do you think we should do?"

Assume you were Rachel. How do you think guests will respond to this competitor's new pricing strategy? What specific steps would you recommend to Ralph that can help Blinky's address this new pricing/cost challenge?

Methods of Food and Beverage Pricing

Many factors impact how a commercial foodservice operation establishes its prices, and the methods used are often as varied as the operators who utilize the methods.

Menu item prices can be directly affected by one or more of the factors previously described. However, in most cases, menu prices have historically been determined based on either an operation's costs or its desired profit level. This makes sense when the operator's perspective of the price formula introduced earlier in this chapter is re-examined closely. That formula was presented as:

Item cost + Desired profit = Selling price

When foodservice operators focus on item costs to establish prices they recognize that items that cost more to produce must be sold at prices higher than lower-cost items. The actual prices charged in a foodservice operation are primarily determined by its owner, perhaps with the assistance of back-of-house production staff. However, those responsible for marketing a foodservice operation should have a good understanding of both the cost-based and the profit-based approaches to pricing menu items.

Cost-Based Pricing

When using cost-based pricing, a foodservice operator calculates the cost of the ingredients required to produce the menu item being sold. The cost of the menu item will also include any menu item accompaniments. This is the case, for example, when a foodservice operation sells a dinner entree that includes a salad, and dinner rolls. In this example, the cost of the entree must also include the cost of the salad, dressing, dinner rolls, and butter that accompanies the service of the dinner rolls.

In the majority of cases, cost-based pricing is based on the idea that the cost of producing an item should be a predetermined percentage of the item's selling price. Under this system, menu items that have lower food costs when compared to their selling prices are typically considered to be more desirable than menu items with higher food costs.

The formula used to compute an item's actual **food cost percentage** is:

$$\frac{\text{Item food cost}}{\text{Selling price}} = \text{Food cost percentage}$$

To illustrate, if it costs an operator \$5.00 to purchase the ingredients needed for a menu item, and the item is sold for \$20.00, the item's food cost percentage would be calculated as:

$$\frac{\$5.00 \text{ Item food cost}}{\$20.00 \text{ Selling price}} = 0.25 \text{ or } 25\%$$

Key Term

Food cost percentage: A ratio calculated by determining the food cost for a menu item and dividing that cost by the selling price of the item.

When operators utilize food cost percentage as a major factor in pricing menu items, it is essential that the cost of producing one portion of the menu item is accurately determined. When the item sold is a single portion item (for example, one New York strip steak), the cost of producing one portion is relatively straightforward.

However, when menu items are produced in multiple portions, for example, when a pan of lasagna containing 12 portions is prepared, operators must calculate their **portion costs** based on the **standardized recipe** used to produce the menu item.

Key Term

Portion cost: The product cost required to produce one serving of a menu item.

The use of standardized recipes is critical for those operators utilizing product cost as a primary determinant of their menu prices. The cost of producing a menu item must be known, and it must be consistently the same if selling prices

Key Term

Standardized recipe: The ingredients needed, and the procedures used to produce a specific menu item.

will be based on product costs. When ingredient costs change, these increased costs must be utilized to calculate the new portion costs that result from the use of standardized recipes made with higher-cost ingredients.

Technology at Work
Foodservice operators today have a wide variety of choices when selecting software programs designed to help them easily calculate the cost of producing their various menu items.

> **Technology at Work**
>
> Foodservice operators today have a wide variety of choices when selecting software programs designed to help them easily calculate the cost of producing their various menu items.
>
> These menu management programs allow operators to insert their standardized recipes and the cost of the ingredients used to make the recipes, and then the items' portion costs are automatically calculated.
>
> Advances in menu management software continue to occur rapidly. Increasingly, foodservice operators are looking for programs that will give them many options to choose from when designing their own pricing and sales tracking processes.
>
> To stay current with newly developed menu pricing software and apps, enter "menu item pricing software" in your favorite search engine, and then review the results.

While commercial foodservice operations are concerned with food cost percentages, noncommercial operations are typically more interested in how much it costs to serve each of their guests.

Operations in which the average cost of meals served is important include military bases, hospitals, senior living facilities, school and college foodservice operations, and business organizations that provide meals to their workers. Whether the guests served are soldiers, patients, residents, students, or employees, calculating an operation's cost per meal is easy because it uses a variation of the basic food cost percentage formula. The formula used to calculate the average cost of meals served in an operation is:

$$\frac{\text{Cost of food sold}}{\text{Total meals served}} = \text{Cost per meal}$$

For example, assume a noncommercial operation incurred $55,000 in cost of food sold during an accounting period. In that same accounting period, the operation served 10,000 meals. To calculate this operation's cost per meal, the cost-per-meal formula is applied.

In this example, it would be:

$$\frac{\$55,000 \text{ Cost of food sold}}{10,000 \text{ Meals served}} = \$5.50 \text{ per meal}$$

Whether managers are most interested in their cost of food percentage or their cost per meal served, it is essential that they first accurately calculate their cost of food sold. When foodservice operators have established a target food (or beverage) cost percentage, they can then use that target to set their menu prices.

For example, if a menu item costs $7.00 to make, and an operation's desired cost percentage is 40% (the percentage is typically the food cost percentage in the approved operating budget for the time period the menu is used), the following formula can determine the item's menu price:

$$\frac{\text{Food cost of the menu item}}{\text{Desired food cost}\%} = \text{Selling price}$$

In this example:

$$\frac{\$7.00 \text{ Food cost of the menu item}}{.40 \text{ Desired food cost}\%} = \$17.50 \text{ Selling price}$$

A second method of calculating selling prices based on predetermined product cost percentage goals can also be used. This method uses a pricing factor, or multiplier, assigned to each potentially desired food or beverage cost percentage. This factor, when multiplied by an item's portion cost, will result in a selling price that yields the desired product cost percentage. Some of the most used pricing factors are presented in Figure 6.2.

DESIRED PRODUCT COST %	PRICING FACTOR
20	5.000
23	4.348
25	4.000
28	3.571
30	3.333
33⅓	3.000
35	2.857
38	2.632
40	2.500
43	2.326
45	2.222

Figure 6.2 Pricing Factor Table

The pricing factor method of establishing menu prices is easy to use. For example, if an operator wanted to achieve a 25% product cost, and a menu item has a food cost of $4.50, the following pricing formula would be used:

Item food cost ($\times$) Pricing factor = Selling price

In this example, that would be:

$4.50 Item food cost ($\times$) 4.0 Pricing factor = $18.00 Selling price

The two methods of arriving at a proposed selling price based on product cost percentage yield identical results. With either approach, the selling price of an item is determined with the goal of achieving a specified food or beverage cost percentage for each item sold.

Technology at Work

Foodservice professionals will have no difficulty identifying a large number of publications detailing a variety of cost-based menu pricing techniques and strategies.

To find a list of currently released publications, go to the Amazon website, select "Books" from the pull-down menu, and then enter either *Menu Pricing* or *Food Cost Control* to review the most recently published works related to the determination of food and beverage menu prices.

One of the best in the market is *Food and Beverage Cost Control*, published by John Wiley, and written by Dr. Lea Dopson and Dr. David Hayes. When you are on the Amazon site, review the index of the latest edition of this popular text.

Contribution Margin-Based Pricing

Some foodservice operators use a more profit-based pricing method, rather than a product cost-based method, to establish menu item selling prices. These operators set their prices based on each of their menu item's **contribution margins (CMs)**.

CM is defined as the amount of money that remains after the product cost of a menu item is subtracted from the item's selling price. It is, then, the amount that a menu item "contributes" to pay for labor and all other expenses and to contribute to the operator's profit margin.

Key Term

Contribution margin (CM): The amount of revenue that remains after a menu item's food cost is subtracted from its selling price.

To illustrate the use of CM pricing, assume a menu item sells for $18.75 and the cost of food cost to produce the item is $7.00. In this example, the CM for the menu item is calculated as:

Selling Price − Item Food Cost = Contribution Margin (CM)

or

$18.75 selling price − $7.00 item food cost = $11.75 CM

When this approach is used, the formula for determining the selling price is:

Item Food Cost + Contribution Margin (CM) Desired = Selling Price

When using the CM approach to establish selling prices, operators develop different CM targets for various menu items or groups of items. For example, in an operation where items are priced separately, entrées might be priced with a CM of $9.50 each, desserts with a CM of $4.25, and nonalcoholic drinks with a CM of $2.75.

To apply the CM method of pricing, foodservice operators utilize a two-step process.

Step 1: Determine average contribution margin required
Step 2: Add the contribution margin required to the item's product cost

Step 1: Operators determine the average CM they require based on the number of items to be sold or on the number of guests to be served. The process used for each approach is identical.

For example, to calculate CM based on the number of items to be sold, operators add their nonfood operating costs to the amount of profit they desire, and then divide the result by the number of items expected to be sold:

$$\frac{\text{Nonfood costs} + \text{profit desired}}{\text{Number of items to be sold}} = \text{CM desired per item}$$

To calculate CM based on the number of guests to be served, operators divide all of their nonfood operating costs, plus the amount of profit they desire, by the number of expected guests:

$$\frac{\text{Nonfood costs} + \text{profit desired}}{\text{Number of guests to be served}} = \text{CM desired per guest served}$$

For example, if an operator's budgeted nonfood operating costs for an accounting period are $125,000, desired profit is $15,000, and the number of items estimated to be sold is 25,000, the operator's desired average CM per item would be calculated as:

$$\frac{\$125,000 \text{ (nonfood costs)} + \$15,000 \text{ (profit desired)}}{25,000 \text{ (Number of items to be sold)}} = \$5.60 \text{ CM desired per item}$$

Step 2: Operators complete this step by adding their desired CM per item (or guest) to the cost of making a menu item. For example, if, as in the example above, an operator's desired average CM per item is $5.60 and a specific menu item's food cost is $3.40, the item's selling price would be calculated as:

$5.60 CM desired + $3.40 item food cost = $9.00 selling price

The CM method of pricing is popular because it is easy to use, and it helps ensure each menu item sold contributes to an operation's profits. When using CM to set menu prices, the prices charged for menu items vary only due to variations in product cost. When managers have accurate budget information about their nonfood costs and realistic profit expectations, the use of the CM method of pricing can be very effective.

Operators who utilize the CM approach to pricing do so in the belief that the average CM per item they sell is a more important consideration in pricing decisions than the product cost percentage. The debate over the "best" pricing method for food and beverage products is likely to continue. However, those responsible for marketing a foodservice operation must remember to view pricing as an important process with the goal of determining a desirable price/value relationship for guests. In the final analysis, the customer will eventually determine what an operation's sales will be on any given menu item. Experienced operators know that sensitivity to required profit as well as to guests' needs, wants, and desires is the most critical component of an effective pricing strategy.

Evaluation of Pricing Efforts

Regardless of the method used to establish selling prices, foodservice operators should regularly evaluate the results of their pricing efforts. While there are several ways to do this, many operations examine each menu item's popularity and profitability. **Menu engineering** is a term popularly used to describe one method that addresses these two variables.

Menu Engineering

Operators using the menu engineering process use it to produce a menu that maximizes the

Key Term

Menu engineering: A system used to evaluate menu pricing and design to create a more profitable menu. It involves categorizing menu items into one of four categories based on the profitability and popularity of each item.

menu's overall CM (defined earlier as the amount operators have available to pay for labor and all other expenses and to contribute to profit). As a result, operators should be keenly focused on their menu's overall CM.

To use menu engineering, operators must sort their menu items by two variables:

1) Popularity (number of each item sold)
2) Weighted contribution margin

Calculating Popularity (Number Sold)

To calculate the average popularity (number sold) of a menu item, operators use the following formula:

$$\frac{\text{Total Number of Menu Items Sold}}{\text{Number of Menu Items Available}} = \text{Average Number Sold}$$

For example, if an operator sold 5000 entrees in a specific time period, and the operator's menu lists 10 entree choices, the average popularity of entrees sold in the time period would be calculated as:

$$\frac{5000 \text{ Menu Items Sold}}{10 \text{ Menu Items Available}} = 500 \text{ Menu Item Average Popularity (Number Sold)}$$

When using menu engineering and applying the 500 average popularity criteria, any menu item that sold more than 500 times in the time period would be classified as "High" in popularity, and those menu items that sold less than 500 times would be classified as "Low" in popularity.

Calculating Weighted Contribution Margin

To continue the menu engineering process, operators must also find the weighted average CM of their menu items. Some operators confuse averages (means) with weighted averages. However, the distinction between the two is important to understand.

To use a simple example, assume an operator collected the following data and wanted to calculate the average size of the sale made in their operation over the 3-day reporting period.

Week Day	Guests Served	Total Sales	Average Sales
Monday	50	$ 500	$10.00
Tuesday	150	$1,650	$11.00
Wednesday	250	$3,000	$12.00
Total/Average	450	$5,150	?

In this example, if the operator wanted to calculate the size of the "average" sale on Monday through Wednesday, they could NOT use the unweighted formula typically used to calculate a mean (average). That *unweighted formula* would be:

$$\frac{\$10.00 + \$11.00 + \$12.00}{3 \text{ days}} = \$11.00 \text{ per day average}$$

In fact, what this operator really wants to know when calculating the average sales for the three days is "*How much did the average guest spend in my operation from Monday through Wednesday?*"

Since the number of guests served each day varied, to properly answer that question, the operator must use a *weighted* average sale formula as shown below:

$$\frac{\$5,150 \text{ (total sales during all 3 days)}}{450 \text{ (total guests served in all 3 days)}} = \$11.44 \text{ weighted average sale per guest}$$

Note that the operator's average sale size resulting from using the unweighted and the weighted average formulas in this example differ.

Returning to menu engineering, to calculate the average **weighted contribution margin** for their menu items, operators must first calculate the contribution margin supplied by all items sold, and then divide it by the number of items sold as shown in Figure 6.3.

Column A in Figure 6.3 lists the name of individual menu items on the menu. Column B lists the number of sales of each of the individual menu items. Note that the sales (popularity) of the items vary from a low of 190 sold (Item 10) to a high of 1050 sold (Item 4). Column C lists the individual CM of each of the 10 items offered for sale, and column D lists the total item CM generated by each menu item.

Key Term

Weighted contribution margin: The contribution margin provided by all menu items divided by the number of items sold. Weighted contribution margin is calculated as:

Total Contribution Margin
of All Items Sold
———————————————
Number of Items Sold

= Weighted contribution margin

The value in Column D is calculated by multiplying the value in Column B times the value in Column C. In this example, the average number sold is 500, and the average weighted CM is $12.43 ($62,174 total item CM / 5000 sold = $12.43).

After an operator has calculated the popularity and the weighted contribution margin of the items listed on the menu, the items are sorted into a 2 × 2 menu engineering matrix containing four squares as shown in Figure 6.4.

Column A	Column B	Column C	Column D
Menu Item	Number Sold	Single Item Contribution Margin	Total Item Contribution Margin
1	250	$ 14.50	$ 3,625
2	250	$ 7.50	$ 1,875
3	525	$ 12.50	$ 6,563
4	1050	$ 17.25	$18,113
5	510	$ 7.00	$ 3,570
6	625	$ 13.50	$ 8,438
7	400	$ 12.75	$ 5,100
8	825	$ 10.50	$ 8,663
9	375	$ 8.25	$ 3,094
10	190	$ 16.50	$ 3,135
Total	5000	$120.25	$62,174
Average (Mean)	500	$ 12.03	
Weighted contribution margin			$ 12.43

Figure 6.3   Total Contribution Margin Worksheet for 10-Item Menu

Figure 6.4 shows that those menu items whose sales are above the average level of popularity (500 sold in this example) are considered to be "High" in popularity, and items that sold less than 500 times are considered "Low" in popularity. Similarly, those menu items whose contribution margins are above the weighted contribution margin average ($12.43 in this example) are considered to be "High" in contribution margin, and those items with a lower contribution margin are considered to be "Low" in contribution margin.

		Popularity	
		Low	High
Contribution Margin	High	High contribution margin Low popularity PUZZLE	High contribution margin High popularity STAR
	Low	Low contribution margin Low popularity DOG	Low contribution margin High popularity PLOW HORSE

Figure 6.4   Menu Engineering Matrix

Many users of menu engineering name the items contained in the four squares for ease of remembering the characteristics of the individual menu items. These commonly used names (Puzzles, Stars, Dogs, and Plow Horses) are also shown in Figure 6.4.

Figure 6.5 shows where each of the ten example menu items listed in Figure 6.3 are located.

		Popularity	
		Low	High
Contribution Margin	High	PUZZLE Menu items 1, 7, and 10	STAR Menu items 3, 4, and 6
	Low	DOG Menu items 2 and 9	PLOW HORSE Menu items 5 and 8

Figure 6.5 Menu Engineering Results

Menu Modifications

Operators should analyze their menus to make modifications and improvements. When using menu engineering to assist in that process, each of the menu items that fall in the four squares produced requires a special marketing strategy. Examples of these suggested menu modification strategies resulting from menu engineering are summarized in Figure 6.6.

Item	Characteristics	Problem	Marketing Strategy
Puzzle	High contribution margin, Low popularity	Marginal due to lack of sales	a) Relocate on menu for greater visibility. b) Consider reducing the selling price.
Star	High contribution margin, High popularity	None	a) Promote well. b) Increase prominence on the menu.
Dog	Low contribution margin, Low popularity	Marginal due to both low contribution margin and lack of sales	a) Remove from the menu. b) Consider offering as a special occasion, but at a higher menu price.
Plow Horse	Low contribution margin, High popularity	Marginal due to low contribution margin	a) Increase the menu price. b) Reduce prominence on the menu. c) Consider reducing portion size.

Figure 6.6 Potential Menu Modifications

Technology at Work

A foodservice operation's menu is much more than a list of foods and beverages. It can be a powerful marketing tool when used to familiarize customers with a brand. It can also get them excited about the unique items an operation offers for sale.

Regularly evaluating individual menu items for their popularity and profitability is an important marketing task because it helps identify profit-producing items and other items that perform poorly. When this information is known, menus can be modified to optimize sales and profits.

Fortunately, there are a number of good software programs on the market that help operators perform a menu engineering analysis. To review the features and costs of such programs enter "menu engineering software" in your favorite search engine and view the results.

One good reason to perform a menu engineering analysis is to identify items whose prices must or can be increased to enhance an operation's profitability. Some operators are hesitant to raise prices fearing that their customers will react negatively. Experienced foodservice operators, however, know that price may not be a determining factor when guests determine where to spend their dining-out dollars.

Quality of customer service, cleanliness, friendliness of staff, and uniqueness of menu items offered are often *more* important than price. As a result, the best operators ensure all of these aspects of their operations meet or exceed their guests' expectations. When they do, price increases are acceptable to guests and that can help ensure an operation's long-term profitability can be implemented.

The many ways foodservice operators establish and assess their menu prices may vary, but most operators produce a physical or digital menu for their guests. Guests must be informed about what an operation offers for sale and what it charges for its various menu items. An operation's menu is a powerful marketing tool. The next chapter examines how operators can effectively create and utilize this unique marketing tool to improve sales and optimize guest satisfaction.

What Would You Do? 6.2

"$39.95 — that's over ten dollars more than we charged for it yesterday!" said Shawn, the Dining Room manager at Chez Franco's restaurant.

Shawn was discussing the day's dinner menu with Aimée, the restaurant's executive chef. Aimée had just shown Shawn the daily menu insert his service staff would use that night. On the night's new menu, he noticed that the price of Lemon Red Snapper with Herb Butter, one of the operation's most popular dishes, had increased overnight. Yesterday it sold for $27.95. Today Aimée had priced it at $39.95.

"Tell me about it," replied Aimée, "our seafood supplier really raised the price on our latest delivery. My snapper cost is now up by almost $9.00 per pound.

That's over $4.00 a portion. The supplier said there was a shortage, and he wasn't sure how long it would last. With the new cost of snapper, I needed this price increase to keep our food cost ratio in line.

Shawn wasn't sure that his servers or the guests they would serve that night would be very happy with Aimée's pricing decision. The snapper was a very popular item, and that meant tonight lots of customers would likely notice the price increase.

Assume you were a server at Chez Franco's on the night this new menu price was initiated. Unless you were instructed otherwise, how would you likely respond to a returning guest who questioned the significant price increase on the red snapper item? What do you think Shawn should tell his servers to say in response to guests' anticipated reactions to this menu item's price increase?

Key Terms

Price (noun)

Price (verb)

Profit margin

Consumer rationality

Wagyu Beef

Restaurant row

Bundling

Food cost percentage

Portion cost

Standardized recipe

Contribution

   margin (CM)

Menu engineering

Weighted

   contribution margin

Operator's 10-Point Tactics for Success Checklist

Evaluate your need for, and the current status of, each of the following operational tactics. For those tactics you think are important, but not yet in place, develop an action plan for its implementation including who will be responsible for the tactic's completion and the target date by which it should be completed.

				If Not Done	
Tactic	Don't Agree (Not Done)	Agree (Done)	Agree (Not Done)	Who Is Responsible?	Target Completion Date
1) Operator understands the importance of effective pricing when creating the marketing message.	———	———	———		
2) Operator has carefully considered the difference between an operator's view of price and their guests' view of price.	———	———	———		

Tactic	Don't Agree (Not Done)	Agree (Done)	Agree (Not Done)	If Not Done	
				Who Is Responsible?	Target Completion Date
3) Operator has considered the impact of economic conditions, local competition, level of service, and guest type when establishing menu prices.	————	————	————		
4) Operator has considered the impact of portion size, delivery style, meal period, and location when establishing menu prices.	————	————	————		
5) Operator understands the concept of bundling when establishing menu prices.	————	————	————		
6) Operator has considered the value of utilizing a cost-based approach when establishing menu prices.	————	————	————		
7) Operator has considered the value of utilizing a contribution margin approach when establishing menu prices.	————	————	————		
8) Operator understands the importance of analyzing their menu as a means of better understanding their guests' purchasing preferences.	————	————	————		
9) Operator understands how to use menu engineering as a means of analyzing their menus.	————	————	————		
10) Operator recognizes the importance of carefully implementing any needed menu price adjustments to minimize guest dissatisfaction.	————	————	————		

7

The Menu as a Marketing Tool

What You Will Learn

1) The Importance of Effective Menu Design
2) How to Create a Food Menu
3) How to Create a Beverage Menu
4) How to Create Digital Display Menus
5) Special Concerns for Off-Premise Menus

Operator's Brief

In this chapter, you will learn that a menu is the list of items sold in a food-service operation, and it is also the way information about available items is communicated to guests. An operation's menu is a powerful marketing tool, and it must be created professionally and in a way that compliments the overall marketing message of the operation.

There are different types of menus. Among the most important are the à la carte, table d'hôte, cyclical, and du jour menus. A foodservice operation's menu is best produced when all of the menu designers involved bring their unique perspectives to the menu's content, design, and development.

The development of an effective food menu is a four-step process: (i) identifying menu categories to be used, (ii) selecting individual menu items, (iii) writing menu copy, and (iv) designing the menu. The same four-step process is used to develop an operation's beverage menu(s).

Digital menus are an increasingly important marketing tool for all foodservice operations. Digital menus can be developed to be viewed on-site and on a user's computer or smart device. Not all foodservice operations require an on-site digital menu, but today all foodservice operations must develop menus suitable for viewing on guests' digital display devices because digital menus are important marketing tools.

(Continued)

> The increasing off-premise consumption of menu items requires operators to carefully consider which menu items should be offered on takeaway menus and how they are packaged and delivered to guests. Sometimes selected menu items may even be included or excluded from the takeaway menu based primarily on their ability to be packaged and held with minimal loss of quality.

CHAPTER OUTLINE

The Importance of the Menu
    The Menu as a Marketing Tool
    Types of Menus
    The Menu Development Team
    Legal Aspects of Menu Design
Creating the Food Menu
    Identifying Menu Categories
    Selecting Individual Menu Items
    Writing Menu Copy
    Key Factors in Successful Food Menu Design
Creating the Beverage Menu
    Beer Menus
    Wine Lists
    Spirits Menus
Digital Display Menus
    On-site Digital Menus
    On-user Device Digital Menus
Special Concerns for Off-Premise Menus
    Choosing Takeaway Menu Items
    Packaging Takeaway Menu Items
    Holding Takeaway Menu Items for Pick-up or Delivery

The Importance of the Menu

The word **menu**, like many other foodservice terms, is of French origin. Literally translated, it means "detailed list." Foodservice operators use this definition when deciding the specific items to be available for guest selection and the prices they will charge for the items.

The term "menu" has a second meaning when, for example, a server provides a copy (usually made of paper although digital examples discussed later in this chapter are also becoming more popular). In this second context, "menu"

Key Term

Menu: A French term meaning "detailed list." In common usage it refers to (i) a foodservice operation's available food and beverage products and (ii) how these items are made known to guests.

refers to the ways available items are made known to guests. The focus of this chapter is primarily on this second usage of "menu" because the way foodservice operators communicate their menu lists to guests is of critical importance to their success.

The Menu as a Marketing Tool

The menu, more than any other management tool, directly affects the marketing, financial, and operating success of every foodservice operation. An effective marketing message delivered through the proper channels will attract guests. However, unpopular menu items and those not prepared and presented to guests in a professional manner, will not attract new and retain existing customers.

Menu planners hope to develop a menu that consistently offers quality products and services to their guests. Menus take on an important marketing function when they can subtly deliver this message of quality through creative, well-thought-out, and accurately portrayed listings of the menu items to be served.

Types of Menus

When some foodservice operators consider their menus, they first think about meal periods (for example breakfast, lunch, and dinner menus) or the types of food and beverage items offered for sale (for example, appetizer menus, dessert menus, or cocktail menus). Other operators may relate menus to where items are offered for sale (dining room menus, banquet menus, and take-away menus).

Another good way to think about the menu is to consider how items are priced on it, and the frequency with which they are offered. The four most common examples of this approach are the **à la carte**, **table d'hôte**, **cyclical**, and **du jour** menus.

À la Carte Menu

The term "à la carte" is also a French term. It refers to a meal in which guests select the individual menu items they desire. Selected items are then prepared (or portioned) to order and are priced individually. For example, a guest in a casual-service restaurant that offers an à la carte

Key Term

À la carte (menu): A menu that lists and prices each menu item separately.

Key Term

Table d'hôte (menu): A menu with a pre-selected number of menu items offered for one fixed price. Pronounced "tah-buhlz-doht." Also known as a "prix fixe" (pronounced prē-'fēks) menu.

Key Term

Cyclical (menu): A menu in which items are offered on a repeating (cyclical) basis.

Key Term

Du jour (menu): A menu featuring items and prices that change daily. Pronounced "duh-zhoor."

dinner menu typically orders and pays separately for an appetizer, entrée (with salad and potato or vegetable), and dessert. Similarly, guests in a non-commercial cash cafeteria operation in a college, or business dining operation, typically select their menu choices as they pass through the cafeteria's line and then pay for each selected item separately.

Table d'Hôte Menu

The French term, "table d'hôte," refers to a meal composed of menu items offered at a fixed price. In this case, the guest pays one price for all the items ordered. For example, a guest ordering dinner would receive an appetizer, entrée (with salad and potato or vegetable), and dessert and pay a single fixed price. This would also be the case, for example, for guests in a non-commercial cash cafeteria operation such as a college or business dining operation who selects a "dinner special" composed of an entrée, one or more accompaniments, and, perhaps, dessert and a beverage and then pays a single fixed price. This type of menu is also known as a "prix fixe" (fixed price) menu.

Cyclical Menu

The term "cyclical" refers to the cycle or frequency with which a specific menu is repeated. Cyclical menus are most often used in non-commercial foodservice organizations where, for example, college students may be consuming all or most of their meals on-site. To reduce boredom and to reduce the need for menu planners to frequently plan new menus, a cyclical (cycle) menu might be used.

For example, in a senior citizen's assisted living facility a weekly menu may be created, and then that same menu would be offered to residents every six weeks. In this example the foodservice operation is on a six week cycle. Note: Cycle menus still offer operators the opportunity to vary menus to account for special occasions or holidays.

Du Jour Menu

The French term, "du jour" means "of the day." In other words, a du jour menu changes daily. Some operations, and especially fine-dining alternatives, may offer only a limited number of items daily, and these change based on the quality of the ingredients that the proprietor or chef can procure at the market that day. Numerous foodservice operations, however, may choose to offer "daily specials" such as a soup du jour, seafood du jour, or coffee du jour.

These four menu types are not mutually exclusive. Foodservice operations frequently offer combinations of more than one type. Consider, for example, the casual service restaurant with an à la carte menu that has a du jour (special-of-the-day) salad, soup, and entrée. Another example: the operation with a table d'hôte menu that changes daily (du jour). Similarly, a non-commercial facility such as a hospital can offer a cyclical menu but charge cash (on an à la carte basis) for hospital visitors dining in the cafeteria.

The Menu Development Team

Regardless of the type of menu being developed, a team of foodservice professionals most often develops the best menus. This is because no single person generally has sufficient knowledge of everything needed to produce an outstanding menu.

For example, a chef or other member on the food production team can provide great knowledge about the ingredients contained in menu items and the methods used to best produce them. This information, of course, will be important as the menu is developed. Other foodservice professionals, however, will have expertise in how menus are actually designed, written, and produced. These marketing experts will be just as important for the development of an effective menu as are the operation's food production personnel.

An effective menu development team should include back-of-house food production staff, and it should include at least one member from the front-of-house staff as well. In fact, regardless of the formal titles of those who assist with menu planning, there should always be persons on the team who bring the following perspectives to the task:

1) Those of the operation's owner or manager
2) Those of the guests for whom the menu is written
3) Those of purchasers who must procure the required items dictated by the menu
4) Those of food production staff who must produce items in the quantities and at the quality levels that are required
5) If applicable, those of beverage production staff who must produce the drinks to be served
6) Those of the service personnel who will be responsible for the face-to-face delivery of menu items to guests
7) Those of the readers of the menu. Attending to this perspective will be those responsible for writing, editing, proofing, layout, and menu design.

Of course, one individual could bring multiple perspectives to the task of menu planning. However, a foodservice operator must be careful not to assume that he or she alone can bring all of these perspectives together in the menu planning process. When planning menus it is generally agreed that the more collaboration involved in the development of the menu the better will be the final product.

Legal Aspects of Menu Design

While foodservice operators have a great deal of freedom in how they write and design their menus, that freedom is limited. As indicated in Chapter 4, there are a number of Federal agencies that place specific restrictions on what foodservice operators can state about their menu items.

The best menus effectively communicate with their readers. The accuracy of the information provided is an essential aspect of this effective communication. To comply with accuracy in menu laws, (see Chapter 2) menu makers must truthfully represent the items being served.

Problems with menu accuracy can occur in several ways. Important areas to be considered as menu makers create their physical menus include references to:

✓ *Quantity.* A two-egg omelet must contain two eggs; an eight-ounce steak should weigh a minimum of eight ounces (before cooking).

✓ *Quality.* The term "**prime**," when used to describe a steak, refers to a specific U.S. Department of Agriculture (USDA) grading standard.

U.S. Grade A or U.S. Fancy for vegetables and Grade AA for eggs and butter also indicate quality grades. Only the quality of the products actually used should be indicated on the menu. If it is not possible to ensure that the intended quality grade will always be available, it is best not to list the quality designation on the menu.

Key Term

Prime (beef): The highest quality grade given to beef graded by the USDA. Prime beef is produced from young, well-fed beef cattle and has abundant marbling (the amount of fat interspersed with lean meat).

✓ *Price.* If there are extra charges (for example, for higher-cost liquors used to make drinks) these prices should be identified. If there are service charges (for example, for groups larger than a specified size), these should be indicated.

✓ *Brand names.* If a specific product brand is specified (Coca Cola or Pepsi Cola, for example), this exact brand should be served.

✓ *Product identification.* For example, maple syrup and maple-flavored syrup are not the same. Neither are orange juice and orange drink or mayonnaise and salad dressing.

✓ *Points of origin.* "Pacific" shrimp, for example, cannot be from the Indian Ocean; Idaho potatoes are not from Wisconsin; Lake Michigan whitefish should not be from Lake Erie.

✓ *Preservation methods.* For example, frozen apple juice should not be represented as "fresh," and canned green beans should likewise not be referred to as "fresh."

✓ *Food preparation.* "Made on-site" is different from a **convenience food** product made elsewhere; also, food should be prepared as stated on the menu. For example, a product described as "sautéed in butter" should not be sautéed with margarine.

Key Term

Convenience food: Any food item that is partially, or fully, prepared before it is purchased by a foodservice operation.

✓ *Verbal and visual presentation.* A menu photograph depicting, for example, eight shrimp on a shrimp platter should accurately represent the number of shrimp served; a specialty drink pictured in a fancy stemware glass with a unique garnish should be served in this manner.

✓ *Dietary and nutritional claims.* The laws addressing dietary and nutritional claims on foodservice menus are perhaps the most extensive of all. This is reasonable because dietary claims are important for those who are health conscious, or those who may be allergic or very sensitive to certain foods. Among the menu terms that have been legally defined by the FDA are:

- *Low fat.* The FDA has established that items referred to as "low fat" should have 3 g of fat or less in a serving. However, if a restaurant serves a portion size larger than the FDA's standard portion, a "low-fat" food may contain a correspondingly larger amount of fat.
- *Lite.* This term can describe the taste, color, or texture of a food, or it may indicate that a food's calorie, fat, or sodium content is reduced. The menu must clearly indicate what "lite" means for the specific menu item.
- *Cholesterol-free.* Foods such as meat, poultry, and seafood naturally contain cholesterol. Menu planners should recognize that "cholesterol-free" does not mean the same as "fat free."
- *Sugar-free.* This term does not mean "calorie-free" or "fat-free."
- *Gluten-free.* This term can only be used when a menu item contains no gluten.
- *Healthy.* Foods noted to be "healthy" should be low in total fat and saturated fat, and they should not be high in cholesterol or sodium. There are not, however, limits on the amount of sugar or calories that a "healthy" food may contain.
- *Heart.* Menu claims such as "heart-healthy" or "heart-smart," and the use of heart symbols imply that a food may help reduce the risk of heart disease. Foods to which a heart-related claim has been made should be low in total fat, saturated fat, and cholesterol, and they should not be high in sodium.

After the menu development team has been selected, and members have been instructed about legal requirements related to producing menu items, they can begin to create an operation's food menu.

Creating the Food Menu

As shown in Figure 7.1, the creation of a food menu is a four-step process:

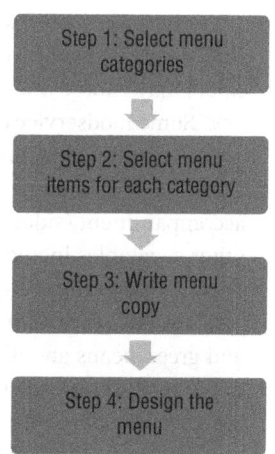

Figure 7.1 The Food Menu Development Process

Identifying Menu Categories

Most menu items are organized into logical categories on the menu. These groupings allow readers to consider choices among reasonably equivalent alternatives. Generally, foodservice operations with higher **guest check averages** tend to offer more menu categories.

For example, a high check average, fine-dining operation may offer numerous appetizers, soups, salads, hot and cold entrées, vegetables, other accompaniments, desserts, and beverages along with a wine list. By contrast, a quick-service property may offer fewer categories, perhaps only sandwiches, side dishes, desserts, and beverages.

Key Term

Guest check average: The average (mean) amount of money spent per guest (or table) during a specific accounting period. Also referred to as "check average" or "ticket average."

In many cases a review of menus from similar operations and a careful analysis of the success of an operation's own menu can provide helpful information about categories to be selected. Typically, foodservice operators identify the following broad categories for potential inclusion on the menu:

Entrées. When reading a menu many foodservice guests choose entrées first and then select other items that go with the selected entrée. Many guests think of entrées as hot items, and many are served hot. However, others, such as a chef's salad, may be cold. Still others may be hot and cold such as a cold Caesar salad topped with a hot grilled chicken or salmon fillet. Regardless of the items included in it, entrées may be the first menu items reviewed by guests so for most operations, this is an essential menu category.

Appetizers. These items, served before the meal to "tempt" one's appetite, are often (but not always) smaller, bite-sized items. They may be hot (such as deep fried pickles) or cold (such as a shrimp cocktail or cheese and fruit). Some light eaters may choose one or more appetizers as their entree.

Soups. Some foodservice operations list soups as a separate category on the menu, and others include them as part of the appetizer category or with salads.

Salads. Many operations offer two types of salads. Entrée salads (see above) and accompaniment (side) salads may be made from lettuce and other greens and other vegetables including coleslaw and potato salad. Salads can also be made of fruits including oranges, melons, pears, apples, and pineapples.

Vegetables and other accompaniments. Vegetables including potatoes, asparagus, and green beans are offered on many menus. Other side dishes such as pastas, dumplings, and noodles are sometimes offered as a menu accompaniment. The specific dish offered is often dictated by the cuisine. For example, in Southwestern and Mexican operations the accompaniments are likely to be rice and beans. In an Italian restaurant, a pasta accompaniment would be more common.

Desserts. After-dinner sweets are often included on the menu with other items. Increasingly, however, a separate menu is used to feature these items. In many cases a dessert menu includes photos of the desserts being offered.

Beverages (non-alcoholic). Many foodservice operations offer guests a wide variety of non-alcoholic beverage choices. Traditional items such as coffees, teas, waters, and soft drinks may be listed on the menu. Sometimes, the brands of soft drinks are also listed.

Foodservice operators must carefully choose the categories to list on their menus. If too few categories are chosen, the menu may be confusing to guests. Choosing too many categories, however, can also be confusing. Remember that an effective menu should list the items available for sale, and also help guests to easily make their selections

Selecting Individual Menu Items

After menu planners have a basic structure in place (desired menu item categories), items to be included in each category must be identified.

Typically, menu items in the entrée category are considered and selected first. Then many menu planners consider, in sequence:

✓ appetizers
✓ potatoes, rice, and other vegetable accompaniments
✓ soups
✓ salads
✓ desserts
✓ beverages

Important factors to consider in selecting menu items include:

✓ *Range (variety).* Normally there should be a range of temperature, preparation methods, texture, shape, and color in the items comprising a menu item category.
✓ *Temperature.* There is typically an expectation that some items will be served hot (meat, poultry, and pork entrées) and other items will be served chilled (dinner and side salads, fruits, and cheeses).
✓ *Preparation method.* Meats can be grilled, fried, braised, or roasted. Vegetables can be boiled, baked, fried, or sautéed, and desserts can be served fresh, chilled, cooked, or baked.
✓ *Texture.* Alternative menu items can be soft, firm, or crunchy. They can be liquid or solid, dull or shiny, and wet or dry. Texture adds to the food's presentation, and it should be considered as items for each menu category are selected.
✓ *Flavor.* While menu planners have traditionally thought of flavor in terms of the basics (sweet, sour, salty, bitter, and the like), today's menu planners include additional flavors such as various degrees of hot (buffalo wings), spicy (Thai

and Cajun dishes), and smoked (barbecue meats and poultry). While the concept of "taste" is very complex and involves more than flavor, guests' perceptions about "what the food should taste like" is very important. When possible, menu makers should include a variety of flavors in each menu item category.

✓ *Color.* Color is an integral part of a menu item's eye appeal. Color may be impacted by preparation method (example: the coating of fried chicken makes a brown product, and broiling the same chicken produces a white item). Good menu planners also recognize that the use of garnishes can help to assure that a menu item's color looks "right" on the plate.

Writing Menu Copy

Menu copy (see Chapter 3) is the term foodservice operators use to indicate the individual descriptions of items listed on a menu. Well-designed menus effectively communicate with guests. As a result, the menu copy used for an item should address both what is being sold and answer basic questions that a menu reader may have. When producing menu copy menu designers should:

✓ *Write plainly.* The first responsibility of well-written menu copy is to describe menu items sufficiently so guests know what they will be buying. Unless the menu is for an ethnic-themed operation,foreign terms should be used sparingly. In addition to helping guests choose the items they want to buy; descriptive menu copy also reduces the order taker's time: they will not continually need to explain important details of menu items.

✓ *Tell guests what they should know.* An effective menu provides a sufficient description of an item so there will be few or no "surprises" when guests receive their orders.

✓ *Spell words correctly.* Menu makers must be careful to ensure the menu copy is grammatically correct and error free.

✓ *Write clearly, punctuate properly, and capitalize consistently.* While most menu readers do not expect menu writers to have perfect grammar, obvious errors on menus will be discovered (and pointed out!) by many guests. In addition, poor grammar may cause confusion. For example, in the following illustration, only the menu copy for item number 4 properly describes a sirloin steak "smothered" with onions and accompanied by a baked potato:

1) Sirloin Steak, served with baked potato smothered in onions.
2) Sirloin Steak, served with baked potato and smothered in onions.
3) Sirloin Steak, smothered in onions and baked potato.
4) Sirloin Steak, smothered in onions, and served with baked potato.

Although descriptions 1 through 3 come close to accurately describing the item, the poor grammar and punctuation can make them confusing (or even comical!) to a menu reader.

✓ *Write and re-write; edit and re-edit.* The task of developing a good menu is just as difficult as (or more difficult than!) developing a good recipe. Significant time will be required as the menu evolves through the many required development and writing phases.

The final menu copy draft should be reviewed by the operation's owner, manager, production personnel and by others to help ensure it effectively explains available menu items.

Key Factors in Successful Food Menu Design

The final step in the menu development process is its actual design. This is a complex step that often requires many decisions. Frequently, menu designers place categories on the menu based on the order in which each is normally served. For example, appetizers will be listed before soups and salads, which are listed before entrées to be followed by desserts.

As menus are designed, some operators use sub-headings within menu item categories. For example, if an operation features several seafood, beef, chicken, and pork entrées, sub-headings to identify each of these entrée types can be helpful. If sub-headings are not used, many menu designers list items of a similar type (for example, beef items or poultry items) together in the entrée category rather than list the beef and poultry items in a random manner throughout the entrée list.

As menus are designed, operators must recognize that menus have **prime real estate** areas in which "Stars" (the operation's most popular and highly profitable items) and "Puzzles" (items that are the most profitable) should be located. Note: Menu engineering is fully addressed in Chapter 6, and areas of menus where Stars and Puzzles should be located are shown in Figure 7.2. Experienced operators understand that, over time, the profitability and popularity of menu items may change, and then design changes are required to accommodate these changes.

Key Term

Prime real estate (menu): A phrase used to define the areas on a menu that are most visible to guests and, therefore, which should contain the items menu planners most want to sell: those that are most popular and profitable.

Regardless of their final design choices, an operation's menu should always be easy to read, and it should not appear cluttered. Some experts even suggest that as much as 50% of the surface space on a menu should be left blank. While this may seem like a lot of wasted space, wide borders around the outside edges of the menu and between categories and the items within them help make menus easy to read.

Figure 7.2 Prime Real Estate Areas on Various Menu Styles

Technology at Work

A growing body of research suggests there is a real science to effective menu design. Principles of psychology and buyer behavior are at work when menus are properly developed. To cite just one example, menu design professionals know that placing an expensive item at the top of a menu category make items listed below that item look more reasonably priced.

An effective menu design should communicate the brand, the vision, and the appropriate guest experience offered by the foodservice operation. Most industry experts agree that effectively designed menus directly impact an operation's revenue producing ability and its guest satisfaction levels.

Fortunately, there are many companies that specialize in assisting foodservice operators in designing their menus. To examine some of these companies and learn about the costs associated with this assistance, enter "menu design specialists" in your favorite search engine and review the results.

Creating the Beverage Menu

Many foodservice operations that serve alcoholic beverages offer one or more separate menus for these items. Creating a beverage menu requires the same four steps used when designing the food menu:

Step 1: Select menu categories
Step 2: Select menu items for each category
Step 3: Write menu copy
Step 4: Design the menu

Just as foodservice operators separate their menus into categories, operators who serve alcoholic beverage products may create categories (Step 1) for the items they sell.

Beer Menus

Beer is any fermented alcoholic beverage made from malted grain and flavored with hops. Lagers and ales are the two basic styles of beer, and some operators design their menus to separate their beer products on these two basic styles. For example, an operator will create beer categories that result in listing lager beers on one part of the menu and ales on another part.

Key Term

Beer: An alcoholic beverage made from malted grain and flavored with hops.

Stouts and Porters are additional beer types, and these may also be listed in a sperate category.

Some operators create menus that categorize their beers based on packaging format rather than style. For these operators, beer categories would include **draft beers,** canned beers, and bottled beers. The operator then creates a beer menu with available products listed under these three categories.

Key Term

Draft beer: An unpasteurized beer product sold in kegs; also known as "tap" beer.

Yet another way for operators selling beer to create menu categories is by basing them on point of production. For these operators, beers are categorized as Domestic beers (those made in the United States) and Imported beers.

Due to their rising popularity, some foodservice operations create special category **craft beer** menus. Since these beers sell for a premium price, they offer a particularly effective way to optimize beer sales by using the beer menu to market locally made beers and microbreweries.

Key Term

Craft beer: A beer produced in limited quantities and with limited availability.

Wine Lists

Operators offering **wine** also make important decisions regarding the categories of products to be placed on their **wine lists**.

Many wine classification categories can be considered as operators produce their wine lists. These include:

Key Term

Wine: An alcoholic beverage made from fruit; most typically grapes.

Key Term

Wine list: A foodservice operation's wine menu.

1) Order of consumption

   When wines are categorized on order of consumption, the wine list is typically separated into Appetizer wines, Entree wines, and Dessert wines.

2) Country (and/or state) of origin

   When wines are categorized by point of origin the wine list will have separate sections for each country whose wines are offered. For example, wines from Germany, France, the United States, and Italy would be listed separately. A point of origin wine list may also separate wines into the states from which they were produced, for example, California, New York, Michigan, and Texas.

3) Color

   Wine color is a popular method of wine list creation. When categorizing by color an operation's wine list typically has categories of red, white, and rosé (blush).

4) Grape used for production

   Wines can be made from nearly any type of grape, but those made from some types of grapes are more popular than others. When operators categorize their wine lists based on the grape used to produce the wine product categories may include, for example, Chardonnay, Cabernet Sauvignon, Pinot Noir, Merlot, Zinfandel, and Syrah.

5) Serving format

   Some operators produce their wine lists based on whether it is offered for sale by the glass, the **carafe**, or the bottle.

Key Term

Carafe: A container used for serving wines. A standard carafe holds one standard size (750 ml) bottle of wine. However, carafe size may vary based on an operation's own service preferences.

6) Carbonation

Some operators categorize wines based on whether they are "sparkling" (carbonated) or "still" (non-carbonated)

7) Selling price

When utilizing selling price to plan wine lists, they are typically listed from most expensive to least expensive or from least to most expensive. When using this system to categorize wines operators should remember that prices should be influenced by the type of operation selling them as well as the quality of wines sold.

Generally, each wine list should include some less expensive wines so cost-conscious guests can enjoy wine with their meal. Similarly, a wine list should include some higher quality and higher-priced wines for those customers who enjoy drinking fine wines with their meals.

Operators who sell wine may categorize their selected product offerings using a combination of two or more classifications. For example, wines can be listed by country of origin and price.

Spirit Menus

Spirits are the most potent of alcohol beverages because they are distilled to increase their concentration of alcohol.

Often referred to as "Cocktail" menus, the development of a spirits menu is directly affected by the Alcohol and Tobacco Tax and Trade Bureau (TTB), a division of the U.S. Department of the Treasury. The TTB sets very specific requirements for the labeling of spirit products. These include:

Key Term

Spirit (beverage): An alcoholic beverage produced by the distillation of fermented grains, fruits, vegetables, or sugar.

✓ Brand or trade name
✓ Class: The broad category "distilled spirits" is divided, under standards of identity, into a number of general but defined classes. Examples include "Neutral Spirits or Alcohol" and "Whisky."
✓ Type: Under most general classes of spirits are specific, defined types of distilled spirits. For example, "Vodka" is a specific type of "Neutral Spirits or Alcohol," and "Straight Bourbon Whisky" is a specific type of "Whisky."
✓ Alcohol content in percent by volume (ABV). **Proof** is optional, but it must be paired with the ABV.
✓ Net contents stated in metric units such as "750 ml"

Key Term

Proof (alcoholic beverage): A measure of the alcohol content of an alcoholic beverage. In the United States, alcohol proof is defined as twice the percentage of "alcohol content in percent by volume," or ABV.

✓ Name and address of the distiller for spirits produced in the United States or the importer of foreign spirits.
✓ Country of origin including where the product was made and where it was bottled.
✓ A commodity statement which identifies the type of distillation utilized. This includes the grains used such as wheat or rye whiskey, and whether the product was made using original distillation or redistillation.
✓ A health claims warning.

Foodservice operators may use any of the required labeling information on spirit bottles to create cocktail menus. Spirits are somewhat unique as a menu item because, in many cases, guests will not actually "see" the products they are served prior to their consumption of ordered cocktails. That is, a guest may be able to see that a served beverage "looks" like a "Gin and Tonic," but they may not see the actual type/brand of gin used to make the cocktail. For this reason it is especially important that cocktails be produced using the exact ingredients listed on the spirits menu.

It is also important to recognize that there is a virtually unlimited number of product combinations that can be made as operators create spirit-based drinks, and there is an equally unlimited number of combinations of spirit products that can be made with various fruits, juices, flavored waters, and other ingredients. As a result, operators serving spirit products have an almost unlimited number of choices related to the categories of their spirit product offerings. Of most importance to the marketing effort of a foodservice operation is that spirit menus clearly indicate the spirit products used in the beverages produced.

Find Out More

The popularity of any specific spirit product can increase or decrease over-time. Currently, whiskey, and especially bourbon, consumption is rapidly increasing in the United States and worldwide. Note: "Whisky" refers to whiskies from Scotland, Japan, and Canada. "Whiskey" refers to whiskeys made in the United States and Ireland.

Whiskey is a distilled spirit made from grains like corn and rye and aged in wooden barrels. Bourbon is a type of whiskey, and there are strict rules in place to ensure its quality. Bourbon must be made in the United States, distilled from at least 51% corn, and aged in new oak-charred barrels.

Contributing to the popularity of whiskey in the foodservice industry is the creation of numerous "Whiskey bars" or "Whiskey lounges." In these operations guests can sample a variety of different whiskies in a comfortable environment.

To learn more about the growing popularity of operations that feature whiskey as their primary menu item offering, enter "whiskey bar popularity" in your favorite search engine and view the results.

Technology at Work

The design and layout of spirit (cocktail) menus are among the most diverse of any foodservice menu. In addition to listing a drink's name and its price, spirit menus can include detailed descriptions of drinks and how they are made as well as photographs of them.

In all cases an effective cocktail menu should encourage guests to drink responsibly and be consistent with the overall marketing message of the operation.

To see examples of the creativity that can be involved in developing a spirits menu, enter "cocktail menu templates" in your favorite search engine and view the results.

What Would You Do? 7.1

"All I am saying is we should use one or two sentences to describe each wine we sell. That way our guests will know more about the exact type of wine they are buying and what entrees will go best with the wines they choose," said Peggy, the wine steward at the Gaylord Bistro, a casual restaurant noted for its chef's imaginative entrée preparations. "It might also make their ordering faster."

"And what I am saying" replied Dan, the operation's dining room manger, "is that with the length of our beer menu, and the two-page specialty cocktail menu we already have, if we add even one or two sentences to describe each wine we sell the beverage menu is going to become massive."

Peggy and Dan were talking to Shivansh, the operation's general manager. All were members of the Gaylord Bistro's menu development planning team.

"Well," said Shivansh, "our current wine list includes about 50 different choices. I agree it's a good idea to tell guests as much as possible about what they're ordering, but my initial thought is that maybe this will actually slow down our dining room service as guests will spend a lot of extra time just reading our menus."

Assume you were Shivansh. After giving it some thought, would you likely agree with Peggy that detailed menus speed guest ordering or would you maintain your initial opinion that they will slow service? How can operators balance their guests' desire to know as much as possible about what they are ordering with the practical risk of potentially producing food or beverage menus that are too long and unwieldy to be effective?

Digital Display Menus

In the past, food and beverage menus would most typically be printed on paper and in a format that was intended to be easily used by guests. While these menus are still extremely prevalent, increasingly all foodservice operators also create a **digital menu**.

A digital menu board is actually a complex system of hardware and software used to display dynamic menus on screens read by guests. In addition to the display of menu items and prices, digital menus can also display specials and promotions, pictures, videos, customer reviews, and more. Essentially, there are two main types of digital menus. The first type is intended to be displayed on-site, and another type is intended to be displayed on a user's computer or smart device.

Key Term

Digital menu: An integrated system that uses hardware and software to display an operation's menu on an electronic screen; also commonly referred to as a digital display menu or digital menu board.

On-site Digital Menus

Digital menus designed to be used on-site may be placed in an operation's interior to enable guests to approach a QSR operation's order counter and view a digital menu prior to placing orders. These on-site digital menu boards are typically colorful and include high resolution images of the menu items being served. These menu boards can be easily changed during the day to display, for example, an operation's separate breakfast, lunch, and dinner menus.

Digital menus may also be designed to be viewed from an operation's exterior and is the type most often seen in foodservice operations with drive-thru service. When thy are placed outside, they must be weatherproof to shield the hardware from environmental elements. They must also be properly designed to be easily read in bright sunlight and at night. They must also be constructed to minimize their chance of damage from vandals or careless drivers.

The hardware in a digital menu board consists of display screens and a media player: the device that downloads menu content from the sign's software. The **content management system (CMS)** in a digital menu board allows operators to create the content guests will view, to upload new content and to then change or modify content as needed.

Key Term

Content management system (CMS): Computer software used to load content into a digital menu display system.

Technology at Work

A well-designed content management system (CMS) allows foodservice operators great flexibility in utilizing their digital menus.

Good CMS software makes it easy for operators to schedule menus by time of day so specific menu items appear only during a defined time. They also make it easy to load videos and animations to time- and date-specific promotions on new products and higher-margin impulse items.

In many cases, multi-unit operators can select a CMS that allows them to create location-specific content. Then, for example, an operation located in a suburban area could, at a given time of day, promote different menu items than a similar operation located in an urban area.

Improvements in CMSs are occurring rapidly and continually. To see some features in systems currently available, enter "content management systems for restaurant menus" in your favorite search engine and review the results.

On-user Device Digital Menus

Viewing menus and ordering online is not new, and Pizza restaurants were pioneers in the online ordering space. For example, Pizza Hut first began experimenting with online ordering in 1994, and it launched its first mobile ordering app in 2009.

Today, foodservice guests' use of digital devices to order menu items is increasing steadily. According to a 2022 survey of users, digital ordering represents 28% of all restaurant orders.[1]

Rather than view a paper menu, however, these guests are viewing an operation's menu on their own user device (typically a cell phone or other handheld smart device).

There are significant differences between viewing a paper menu and a menu online. Operators must address these differences when posting menus online, and three of these important differences are page size, font size, and file size.

Page Size

Operators must recognize that a standard 8.5″ × 11″ piece of paper is about three times as wide as the average (4 inches) smartphone screen, and printed menus might be even larger than a standard size piece of paper. Unfortunately, operators who simply take a photo of their printed menus for online posting make it difficult for potential guests to read and navigate (see Chapter 9) these online menus.

---

1 https://blog.trycake.com/online-ordering-stats-to-know-in-2022#:~:text=In%202022%2C%20 the%20number%20of,represents%2028%25%20of%20all%20orders retrieved June 17, 2022.

The better approach is to post menu images online that will automatically scale to fit the width of a mobile phone screen. Unlike print menus, there are no page boundaries or height limitations involved in designing a digital menu for viewing on a smart device. Therefore, operators can stack their content into one column and allow the volume of content to determine the page height. A smart device's screen automatically limits the content seen at one time, so users can scroll through the menu at a comfortable pace in much the same way as when they check their emails or read a news article online.

Font Size

A second important factor to remember is that **font size** best used for paper menus differ from those best used for viewing menus on a user-owned smart device.

Font size is measured in points (pt.) that dictate lettering height. There are approximately 72 (72.27) points in 1 inch (2.54 cm). As a result, for example, a font size letter of 72 pt. would be about 1 inch tall, and a font size letter of 36 pt. would be about one half of an inch tall.

Key Term

Font size: The size of the characters printed on a page or displayed on a screen. Commonly referred to as type size.

Figure 7.3 Font Sizing Examples for Smart Devices

Figure 7.3 shows the way various font sizes would appear on a typical user's 4-inch smart device screen.

Some general font sizing recommendations based on a smart device page width of 4 inches are:

- ✓ Category Headings (e.g., Appetizers, Salads, Entrees, and Desserts) 28–32 pt.
- ✓ Menu item names: 14–20 pt.
- ✓ Menu item descriptions: 12–14 pt.

As with printed menus, type sizes that are too small to easily read on a user's screen can be irritating and can create serious mistakes when guests order. Most importantly, if guests cannot easily read a menu and are continually forced to manually expand their screens to do so, they may simply decide to order from another operation.

File Size

Smart device users download content at various speeds depending on their internet or phone connections and the quality of device being used. It is best for operators to keep menu files as small as possible to optimize them for web viewing. In general, the smaller the file the faster it will load on a smart device. Therefore, operators should have separate menu files for each of their major menu categories. Then, for example, a guest interested in looking at entrees need not wait for dessert content to be loaded before proceeding.

Operators should understand that videos, detailed images (high resolution photos), and complex graphics displayed on a screen require more data download time than text downloads. If they want to show a variety of images such as dining room interiors or menu items, they should select such images carefully and only use them when they significantly add to the viewer's menu reading experience.

Technology at Work

Digital menus viewed on a smart device differ from print menus. When creating a printed menu the amount of information presented should be minimized to reduce the menu's length and the number of required print pages. Since digital menus are not printed, their length is not a significant concern. Operators may continually add information since viewers know they can scroll up and down on web displays to view content.

Alternatively, high resolution photos ideal for use on a print menu may frustrate smart device users because of the download time and data usage involved in downloading and viewing them.

Many operators will likely require assistance to properly convert their print menus to digital menus suitable for smart device viewing. Fortunately, numerous companies specialize in assisting operators in designing their menus for smart devices. To examine some of these companies and learn about the costs associated with this assistance, enter "converting print menu to mobile friendly menus" in your favorite search engine and review the results.

Special Concerns for Off-Premise Menus

An **off-premise menu** is one that foodservice operators produce especially for use by those guests who wish to purchase menu items that will be picked up or delivered to them. An off-premise menu can be identical to an operator's

Key Term

Off-premise menu: A menu listing items available for pick up or delivery; also commonly referred to as a takeaway, take-out, or carry-out menu.

normal menu, but in most cases it should not be. The reasons why are many, but all relate to the challenges inherent in maintaining product quality for off-premise menu item consumption. While not every foodservice operation will choose to produce a separate off-premise menu, there can be significant advantages to doing so.

The COVID-19 pandemic of 2020–2022 impacted foodservice operations in many important ways. Some operations that had never offered menu items for take-out or delivery had to do so to stay in business. In addition, some operations that had previously generated only a relatively small percentage of their revenue from off-premise sales found that these sales represented most or all of their entire revenue. What these operators discovered was that there are special concerns related to selecting, packaging, and holding of menu items to be consumed off-premise.

Choosing Takeaway Menu Items

Menu items to be sold for off-premise consumption must be carefully chosen because the time between order pick-up and item consumption is not known in advance. For guests picking up menu items and transporting them to a nearby location the time between item pick up and consumption may be very short. Alternatively, on a busy night when a guest order has been prepared for pickup and delivery by a third-party delivery service, the length of time items will be held before consumption may be relatively long.

The marketing aspects of menu item selection for take away menu items is important to recognize because guests will always equate an operation's quality standards with the products they receive when they are picked up or delivered.

Some menu items can be held for a relatively long time with no reduction in quality. Examples include pizza, fried chicken, deli sandwiches, and undressed salads. Some menu items, however, simply do not travel well. Examples are many deep-fried items (including French fries!), ice cream-based items, Eggs Benedict, souffles, and delicate fish dishes.

Each foodservice operation is different. Operators should carefully consider whether items that deteriorate if not quickly consumed should be offered on off-premise menus. Troublesome items that do not reflect the quality of the entire menu may then be omitted in favor of other items that can be transported while preserving food quality, texture, and flavor.

Packaging Takeaway Menu Items

The packaging of take away items is just as important as their selection for inclusion on a take away menu.

For example, assume an operator has just received an order for a green salad, burger, French fries, and a cold fountain drink. The order will be picked up and delivered by a third-party delivery partner to a customer who lives several miles away. If this guest's order is not packaged correctly, the salad may be wilted (if dressing were added prior to the salad's packaging), the burger and fries may be cold (if an appropriate hot foods container was not utilized), and the cold fountain drink may contain melted ice and could be spilled during delivery (if an appropriate cold food container was not utilized, and its lid did not fit snugly). In this example guest satisfaction will be negatively impacted not because of the products selected, but because the selected products were improperly packaged.

As a second example, assume a foodservice operation received a pick up order for a three chicken taco meal. The tacos *could* be fully assembled and placed in a Styrofoam container and, if so, the food would technically be ready for pick up. It would be much better, however, if the chicken fillings were packaged in a hot container, chilled items like sour cream, cheese, and shredded lettuce were packaged in a cold container, and the warm tortillas were wrapped in tin foil. In this second packaging scenario, even if the guest's pick up time was delayed, product quality would be optimized.

While each foodservice operation is different, fundamental "good marketing" principles of take away food packaging include:

✓ Packaging hot foods separate from cold foods
✓ Packaging liquid foods in containers with secure fitting lids
✓ Utilizing see through containers whenever possible
✓ Replacing plastic containers with more environmentally friendly containers when possible
✓ Customize take away containers and products with the operation's name or logo where it is economically feasible to do so

---

Find Out More

Take away foods must be packaged properly for food safety and product quality. While many foodservice operations utilize Styrofoam clamshells as their go-to packaging option (and Styrofoam does help keep hot foods hot and cold foods cold), these containers can break open easily and lead to messy leaks at the slightest bit of turbulence. As a result, many operators increasingly look to other packaging options that provide a strong seal to keep all ingredients intact. In addition, most foodservice operators include condiments, appropriate cutlery, and napkins with their take away orders.

Fortunately, foodservice operators can now choose from a number of alternative products that help maintain the quality of take away foods and that are environmentally friendly. Examples of these items include bamboo plates, wood dinnerware, and items made from sugar cane or bagasse (a by-product of sugar cane fiber) dinnerware. Other items include kraft paperboard made from 100% unbleached paperboard with an additional coating that makes the product cut-and-leak-resistant.

To learn more about these and other new packaging options enter "innovative take away food packaging" in your favorite search engine and view the results.

Holding Takeaway Menu Items for Pickup or Delivery

Even when menu items are carefully selected for suitability as takeout items and if they are packaged properly, product quality can still decline. This is especially so if these items are not held properly until they are picked up by guests or by delivery drivers.

Guest pick up of ordered items can be delayed by traffic problems, other unforeseen circumstances causing delays, or simply by miscalculation of travel times. Order pickup by third-party drivers can be affected by issues including the total number of deliveries requested at a specific time (i.e., some times of the day are busier than other times), the number of available drivers, or the amount the guest will tip the driver. All these issues are addressed in detail in Chapter 10.

Regardless of the reasons, however, some takeaway orders will be held on-site for extended periods of time and longer holding times can result in product quality deterioration. To minimize the negative impact on food quality of excessive holding times, operators must ensure that hot food packages are held separately from cold food packages. In many cases, this will require separate containers or bags for a single pickup order. If hot and cold foods must be packaged together in the same bag, box, or container, the order should be assembled upon a guest's or a driver's arrival when possible.

Holding areas for takeaway items should be clean and brightly lit for ease in labeling and packaging orders. For guest orders that will be delivered in coolers or insulated bags, these items should be properly cleaned, sanitized, and disinfected between uses. Increasingly, some operators add tamper-evident labels (those that, when ripped or torn, alert guests that their order was tampered with), or tamper-proof boxes can also be used for their deliveries.

The menu is a vital marketing tool for every foodservice operation. In many cases, it will be the menu items sold that will be featured most prominently in an operation's marketing mix. In Chapter 1, "Marketing Mix" was defined as the specific ways a business utilizes the 4 Ps of marketing to communicate with its potential customers. In the next chapter, we will closely examine "Promotion;" the P of marketing that includes the important topics of advertising, personal selling, promotions, publicity, and public relations.

Technology at Work

As takeout meals and third-party delivery companies continue to grow in popularity, the need for appropriate takeout packaging continues to evolve.

Peace of mind for customers receiving delivered meals is important and as a result, there has been increased demand for tamperproof, or tamper-evident packaging of delivered meals. This type of packaging helps ensure that food has not been "sampled" or otherwise adulterated by delivery drivers.

Tamperproof packaging provides peace of mind both to guests and to food-service operations. Guests can be assured that their delivered menu items are safe and unadulterated. Foodservice operators can utilize third-party delivery services with reduced fear of the tarnishing of their operations' names by unscrupulous delivery drivers.

Innovative packaging for takeout and delivery foods continues to evolve. To learn more about the tamperproof products resulting from this evolution, enter "tamperproof packaging for restaurant delivery" in your favorite search engine and review the results.

What Would You Do? 7.2

"I can't believe this guest!" exclaimed Malachi, as he looked over the shoulder of Rita, the owner of Smiley's Pizza, a ghost kitchen that serves take-out pizza and related Italian-style appetizers and desserts. Malachi, the operation's kitchen manager, and Rita were looking at Rita's smartphone and reading a guest's online review of their operation. The review had just recently been posted on a popular social media site.

"This guest says that their pizza was fine, but the Fried Mozzarella Sticks they ordered with it were cold and soggy when they were delivered," said Rita.

"I was working in the kitchen when the guest placed the order. We were really busy, and I can't remember the exact order, but I do know we didn't have any problems getting all orders out on time," said Malachi.

"Then I guess we need to look at a potential delivery issue," said Rita. "If our third-party delivery partner picked up more than one order from us at the same time that night and, if this guest's order were delivered last, that could explain the problem. Regardless, the bottom line is that this guest only gave us a 1-star rating!"

Assume you were Rita. Do you think the guest cares whose fault it was that their mozzarella sticks were cold and soggy? Do you think this guest's complaint was primarily due to a production failure, a service failure, or a take-out menu design failure? Why do you think so?

Key Terms

Menu	Cyclical (menu)	Convenience food
À la carte (menu)	Du jour (menu)	Guest check average
Table d'hôte (menu)	Prime (beef)	Prime real estate (menu)

Beer	Carafe	Content management
Draft beer	Spirit (beverage)	system (CMS)
Craft beer	Proof (alcoholic	Font size
Wine	beverage)	Off-premise menu
Wine list	Digital menu	

Operator's 10-Point Tactics for Success Checklist

Evaluate your need for, and the current status of, each of the following operational tactics. For those tactics you think are important, but not yet in place, develop an action plan for its implementation including who will be responsible for the tactic's completion and the target date by which it should be completed.

Tactic	Don't Agree (Not Done)	Agree (Done)	Agree (Not Done)	Who Is Responsible?	Target Completion Date
				If Not Done	
1) Operator understands the importance of utilizing the menu as a marketing tool.	____	____	____		
2) Operator has carefully selected a menu development team to assist in the creation of the operation's needed menus.	____	____	____		
3) Operator has reviewed the basic requirements of truth-in-menu laws and how they affect menu development.	____	____	____		
4) Operator recognizes the initial steps in creating a menu are identifying menu categories and selecting individual menu items.	____	____	____		
5) Operator recognizes the final steps in menu development are writing menu copy and properly designing the menu.	____	____	____		

Tactic	Don't Agree (Not Done)	Agree (Done)	Agree (Not Done)	If Not Done	
				Who Is Responsible?	Target Completion Date
6) Operator understands the basic alternatives available when creating beer menus.	___	___	___		
7) Operator understands the basic alternatives available when creating wine lists.	___	___	___		
8) Operator understands the basic alternatives available when creating a spirit menu.	___	___	___		
9) Operator has reviewed and recognizes the challenges related to digital menu development.	___	___	___		
10) Operator has reviewed and selected off-premise menu items based, at least in part, on each item's ability to be packaged, held, and delivered properly.	___	___	___		

8

Importance of the Foodservice Marketing Mix

What You Will Learn

1) The Importance of Advertising in the Marketing Mix
2) The Importance of Personal Selling in the Marketing Mix
3) The Importance of Promotion in the Marketing Mix
4) The Importance of Publicity in the Marketing Mix
5) The Importance of Public Relations in the Marketing Mix

Operator's Brief

In Chapter 1, you learned about the 4 Ps of the marketing mix (Product, Place, Price, and Promotion). You also learned that several activities are involved with "Promotion," and one of these is advertising. The authors provide focused details about the advertising activity in this chapter because it is so frequently used in so many different forms in almost every foodservice operation.

Advertising is paid communication used to inform, convince, or remind. Foodservice operators use a four-step process to develop their advertising messages: Step 1: Identify the target market, Step 2: Focus on target market needs, Step 3: Present a solution, and Step 4: Explain their uniqueness. When these steps are completed, operators next select the communication channels judged best to deliver their advertising messages and the frequency with which messages are delivered.

The other four promotional activities that are part of the marketing mix are also explained in this chapter. Personal selling is a face-to-face interaction between an operation and its guests. Its goals include increasing visibility in the local community, customer counts, and guest check averages. Personal selling occurs on-premises by an operation's guest service staff and off-premises, usually by an operation's owner or manager.

Promotions are undertaken to provide special buying opportunities to guests. They can focus on all guests, new guests, or repeat customers by the use of an effective customer loyalty program.

Publicity is also part of an operation's marketing mix. Since it is free and can be very powerful, all operators should plan events and actions to generate positive publicity in their marketing plans. They must also, however, develop a crisis management plan (CMP) to properly respond to significant negative publicity if it is encountered.

Public relations is the final part of the marketing mix, and it addresses an operation's external image. These types of efforts should include positive interactions with members of the local media, the operation's community, and its own employees.

CHAPTER OUTLINE

Advertising
    The Purpose of Advertising
    Developing the Advertising Message
    Determining the Best Message Delivery Channels
    Optimizing the Frequency of Message Delivery
Personal Selling
    The Goals of In-person Selling
    On-premise Personal Selling
    Off-premise Personal Selling
Promotions
    The Purpose of Promotions
    Promotions for All Customers
    Promotions for Loyal Customers
Publicity
    Initiating Positive Publicity
    Addressing Negative Publicity
Public Relations
    Media Affiliations and Successful Public Relations
    Community Relations and Successful Public Relations
    Employee Relations and Successful Public Relations

Advertising

As shown in Figure 8.1, "Promotion" consists of five unique marketing activities that are addressed in detail in this chapter.

Advertising is the part of promotion that communicates marketing messages to a foodservice

Key Term

Advertising: Any form of marketing message delivery for which a foodservice operation must pay.

Figure 8.1 Five Components of the Promotion Mix

operation's target market(s). Since the operation must pay for advertising, it is essential that all money spent on advertising is used wisely.

The Purpose of Advertising

The purpose of advertising is to deliver a marketing message to an operation's target audience. While specifics of the marketing message vary based on the operation, one good way to understand advertising is to view it as having one of three main purposes:

1) To inform
2) To convince
3) To remind

To Inform

When a new foodservice operation opens, or when an existing operation is relaunched under a new name, the operation's target market must be made aware of the opening. Informing guests that an operation is open for business most often involves utilizing a paid advertising message. In many operations, this paid advertisement is utilized in conjunction with a well-publicized **grand opening**.

Key Term

Grand opening: A well-publicized event that marks the first day of operations for a foodservice business.

Note: Sometimes "grand openings" occur several days after a new foodservice facility opens. This "slow opening" allows the manager and staff to address any "surprise" problems. It also better ensures that from the time of the first guest's arrival at the "grand opening," all operating procedures are performed exactly as they should be.

For a newly opened foodservice operation, advertising informs potential guests about the name of the operation, where it is located, hours of operation, and any other information guests require to visit the operation.

Effective advertising can take the information dissemination process even further. Examples include describing the service style of the operation, types of menu items to be served including any specialty items offered, the unique setting of the operation, and any distinct the operation provides. Advertisements intended to inform guests should, in most cases, include an explanation of the features and benefits the operation offers to customers.

Advertising used to inform guests is important even for those operations that have been in business for many years. A continuing business wants customers to recall its name and to do so before recalling the names of competitors (who will also be using their advertising messages to inform).

Names and places are easily forgotten, and advertising used primarily to inform and remind guests of an operation should be a part of every operation's advertising efforts.

Find Out More

Holding a grand opening for a foodservice operation is very exciting, and it is incredibly important that it be done correctly. When it is, arriving guests will see the best of what an operation can provide to customers. If a grand opening is not planned and executed properly. However, it can be very frustrating for the business's owner and guests who attend.

Designating a day, weekend, or even a week of special events to mark the opening of a business can create significant guest traffic. To see some ideas related to advertising and executing an effective grand opening event, enter "tips for holding a restaurant grand opening" in your favorite browser and view the results.

To Convince

Most foodservice customers have a variety of choices available to them. Therefore, one major purpose of advertising is to convince target customers that an operation's products and services are a *better value* than that of their competitors. After a target market has been informed about an operation, they should learn reasons why they should select the advertising operation rather than another.

Remember that foodservice operations can focus on the quality and uniqueness of their products and/or their service style (see Chapter 3) as they promote their brands and persuade customers to choose them first. In most cases, this portion of an advertising message should appeal to the emotions of a target market.

To Remind

The most successful advertisements remind customers of the benefits of an operation's products and services. If a foodservice operation does not continually do so, it runs the risk of those customers being lured away by competitors.

Successful companies such as McDonald's or Starbucks are not concerned about informing their guests that they exist or convincing their guests that they are a good choice. Rather, their reminder-related advertisements encourage guests to recall the good experiences they have received in the past, and to encourage them to return in the future.

Foodservice operators should recognize that effective advertising can encourage customers to visit an operation. However, guests who visit because of effective advertising must receive the promised quality of products and services. If they receive substandard products and services, they will quickly become dissatisfied.

Developing the Advertising Message

Chapter 4 examined the use of the AIDA (Attention, Interest, Desire, and Action) guide to developing a marketing message, and it is a good one for determining messages in an operation's **advertising campaign**.

The advertising messages created as part of an advertising campaign can vary as can the text, visual, auditory, or video communications designed to persuade its target market to perform a desired action.

Determining the advertising message is just as important as deciding how to deliver the message. One good way for operators to create their advertising message is to follow a four-step plan as shown in Figure 8.2.

Key Term

Advertising campaign: A specifically designed strategy conducted across different communication channels to achieve specific results including increased brand awareness, higher sales, and improved communication with a specific target market.

Figure 8.2 Four-Step Advertising Message Development Process

Step 1: Identify the Target Market

The first step in developing an advertising message is to identify the target market and then focus the message on that specific market. Every foodservice operation has a target market, even if they are not aware of it. For example, a local pizza parlor operator may feel that anyone who wants a pizza is their target market. However, the target market for this operation is likely limited to those who live or work within four or five miles of the operation. Target market identification is critical to establishing an effective advertising message.

Step 2: Focus on Target Market Needs

Recall that all product or service buyers are seeking to satisfy one or more needs. While they vary, an effective advertising message should directly consider one or more of these needs by addressing the problems to be solved and/or the needs to be satisfied.

When developing an advertising message that focuses on target market needs, it is important to recall the old business adage that "people don't care about you until they know you care." The best advertising messages show that the advertiser understands the target market's problems and recognizes what they want or need.

Step 3: Present a Solution

After a target market's needs and/or desires have been identified, the advertisement should next identify and clearly state the solution offered by the foodservice operation to fully address these challenges. This is the part of the advertising message where the foodservice operator says, "I know what you are looking for, and that is exactly what I can provide for you!"

Step 4: Explain Uniqueness

In nearly all situations, the foodservice business is competitive. Therefore, those in target markets have choices as they make decisions about the operations to visit and from whom they should buy. The operator must specifically explain what is offered, why that solution is unique, and the reasons why the suggested solutions are the best for the customer.

Those unique solutions may be based on several features including product offered, service style, location, or price, among many others. Regardless of the business characteristic being featured, an effective advertising message convinces listeners or viewers that, despite alternatives that may be available, their needs are *best* satisfied by the offerings of the advertiser.

One mistake made by some foodservice operators when crafting their advertising message is the focus on what foodservice operations "do" rather than on what target markets "need." All foodservice operations, by definition, offer food to their guests. The best advertisements focus on guests and what they will receive when they visit an operation (and not on the operation itself).

Determining the Best Message Delivery Channel

Recall from Chapter 3 that a marketing (communication) channel is the form or method used to deliver a marketing message. Sometimes an operation's advertising message dictates the best channel for the delivery of the message. For example, a foodservice operator wanting to *inform* potential guests of its newly renovated dining room would not use radio because this marketing channel radio could not show photos or images of the newly decorated area. In this example, a marketing channel using video or color images would be a better choice.

Similarly, if an operator's advertising message is to *remind* a target market that a product or service is available, that message is not trying to inform or persuade an audience. In this example, a display ad on Google or Facebook may be an excellent option.

Different message delivery channels require varying levels of management time and effort, as well as different levels of marketing funds. Foodservice operators who have carefully crafted their advertising messages must just as carefully choose the communication channels they will use to deliver those messages.

Optimizing the Frequency of Message Delivery

To best optimize the frequency of delivering an advertising message, foodservice operators must recognize the difference between advertising **reach** and **frequency**.

Reach measures the total number of viewers or listeners during a specific time that *could* be exposed to an advertisement. For example, if a local radio station has 100,000 listeners who regularly listen to a specific radio program broadcast at a particular time of day, the reach of an ad broadcast during that show would be 100,000 listeners.

Key Term

Reach (advertising): The total number of viewers or listeners who could be exposed to an ad in a specific time frame.

Key Term

Frequency (advertising): The total number of times a specific viewer or listener is exposed to an advertisement.

Frequency refers to the number of times a specific individual might be exposed to an ad. Frequency is measured in days, weeks, or months. In the previous example, if the radio advertisement were broadcast once per day for two weeks, the frequency would be 14 for the two-week period.

The optimal frequency for an advertisement most often depends on whether an operation is running a long-term or short-term ad campaign. Longer-term campaigns should utilize lower frequency strategies because viewers will be exposed to the ad over a longer time period. Short-term message campaigns should utilize increased frequency for the length of the campaign.

Whether one uses traditional or web-based communication channels (see Chapter 4) to deliver intended messages, the number of advertising channels used to run an ad also affects frequency. The reason: additional channels utilized should increase an advertisement's reach, and its frequency can most often be reduced.

Technology at Work

The Internet continues to open new communication opportunities for both small and large foodservice operators. This is especially so with social media sites. Popular social media sites help operators reach prospective customers, learn about their needs, convey their advertising messages, and obtain instant feedback.

With the aid of available social media monitoring tools, foodservice operators can now track, listen, and gather essential information concerning their businesses and customers across multiple social media platforms.

However, with 250-plus social media sites now operating, it can be difficult to choose the right ones for advertising. Fortunately, many companies now provide services that assist foodservice operators in selecting social media sites that offer maximum advertising impact.

To view social media tracking organizations that can assist foodservice operators in determining the effectiveness of current ads, and to recommend placement of future ads, enter "social media tracking assistance for restaurants" in your favorite search engine and view the results.

Personal Selling

Most foodservice operations are too small to have an individual assigned full-time to the task of **personal selling**.

In some cases, however (examples: facilities that do significant amounts of catering and those who host many large events), there may be an individual designated as a full- or part-time external salesperson. All foodservice operations, however, can benefit from understanding the power of personal selling and how to utilize it.

Key Term

Personal selling: Face-to-face selling-focused interactions that take place between foodservice personnel and their customers.

The Goals of In-Person Selling

Important goals of in-person selling include increased customer counts, optimized operational visibility in the local community, and increased per-guest sales.

Every foodservice operation should focus on increasing guest counts even if it is continually busy because competitive operations will continually advertise their products and services to increase their own market share. When a market is not growing or doing so slowly, the only way to increase market share is to take it from other operations. Therefore, experienced foodservice operators know that a continued effort to maintain and increase current customer accounts is essential for growth. In-person selling is one marketing method that can assist in increasing customer counts.

Operators may increase customer counts by looking for volume business opportunities. Even the smallest foodservice operation might use personal selling efforts to advertise their ability to host special events including holiday and anniversary parties, birthday events, and off-site catering opportunities.

In addition to building volume, in-person selling enables foodservice operations to increase visibility in the local community. For example, if a foodservice operator is active in the local Chamber of Commerce, their activities with that organization will likely be closely associated with their business. The result: goodwill for the business and potential increases in guest counts because of greater visibility in the business community.

Experienced foodservice operators know that volunteering their time in local community's charitable events, supporting worthy local causes, and becoming active in their communities are all excellent opportunities to utilize in-person selling to promote their operational visibility.

When evaluating the impacts of personal selling, remember that effective in-person selling is different from pressuring potential buyers to purchase something not needed or wanted. The best in-person salespeople offer advice and information while carefully explaining what their own organizations can do to meet a potential guest's needs. When they do that effectively, in-person selling can contribute to a strong trusting relationship between the buyer and the seller.

It is also advantageous to increase the per-guest sale (i.e., guest check average; see Chapter 7), and in-person on-premise selling is an effective way to do so. However, in-person selling efforts to increase an operation's guest check average are typically undertaken by an operation's staff rather than its owners or operators. Therefore, training staff in effective in-person selling activities is essential. Operators should recognize that the seating capacity of many foodservice operations is limited and, if so, they should undertake efforts to ensure the per-visit spending level of each guest is optimized.

Regardless of its intended goal, foodservice operators should consider in-person selling efforts that are performed both on- and off-premise.

On-Premise Personal Selling

All guest service employees in a foodservice operation should be part of an operation's on-premise personal selling efforts. Effective on-premise personal selling begins with service employees who have been trained and who are knowledgeable about the menu items being sold, their preparation styles, and their selling prices. When guest service employees are well-trained, they can assist with **up-selling** efforts.

Key Term

Up-selling: The process of increasing the guest check average through the effective use of on-premise selling techniques; also referred to as "suggestive selling."

Effective up-selling training must focus on what the menu items (and services!) an operation sells. Some operators do not train service staff in up-selling because they are concerned about up-setting guests who do not want the menu items being offered. However, that is never the goal of up-selling. Instead, up-selling is best viewed as a process of informing guests about menu choices so their personal dining experiences can be optimized.

For example, if a guest orders a medium rare sirloin steak, it makes good sense for the guest's server to ask if they would like the steak topped with sauteed mushrooms or the operation's house-made "Peppercorn Sauce" made with cognac, whole peppercorns, broth, and heavy cream.

In this example, the guest is simply being informed about available enhancements to the ordered steak, and the guest can either select or not order them. If the guest chooses to enhance their steak with a high-quality topping, the operation's guest check average will be increased and, most likely, the server's tip will also be increased. Up-selling in a foodservice operation benefits everyone and, as seen in this example, sharing important product knowledge about menu items available can benefit the guest, the operation, and the server.

Up-selling opportunities vary based on an operation's menu. The operator should identify the best opportunities to do so and provide service staff with the required product knowledge to effectively up-sell. Note: It is also important to recognize that understaffing is the enemy of up-selling. When servers do not have time to communicate effectively with their guests, up-selling opportunities will be lost. This is one reason it is important that guest service teams are fully staffed during an operation's busiest times.

Find Out More

One key requirement for effective up-selling is a team of well-trained servers who can up-sell without annoying guests. Fortunately, there is much information available on the Internet about how best to train servers in up-selling.

(Continued)

In some cases, implementing an effective up-selling strategy is very simple. Nearly everyone is familiar with the question made famous at McDonald's; "Would you like fries with that?" and this is an excellent example of effective up-selling. The question is straightforward, is not a high-pressure sales tactic, and it leaves the guest completely free to make choices (and, of course, asking the question sells more French fries!)

To see additional ideas related to training servers in effective up-selling techniques, enter "tips for up-selling in restaurants" in your favorite browser and view the results.

Off-Premise Personal Selling

Some foodservice operators set aside a certain amount of time per week to make off-premise **sales calls**. These sales calls simply involve meeting with potential customers who are current customers and/or those who may become customers in the future.

Key Term

Sales call: A conversation between a salesperson and a prospective customer about the purchase of a product or service.

The primary purpose of off-premise sales calls is to keep the foodservice operation foremost in the minds of potential customers. For example, every foodservice operation has product suppliers, and these businesses may seek locations for parties to recognize employee anniversaries or even meetings of their own staff. A sales call to these suppliers emphasizing that the operation has on-site meeting facilities or does catering can be helpful when the supplier plans special events.

Similarly, there may be numerous civic, school-related, religious, and other organizations desiring meals as part of events that can be offered in the foodservice facility. There is an old saying in sales that "People prefer to do business with people they know." For foodservice operators, becoming known in their communities can ultimately result in greater sales.

Foodservice operators' sales calls need not be high-pressure events. Instead, they can involve making an appointment, sharing the operation's menu, pricing policies, and operating hours, and ending the sales call with contact information. When performed effectively, off-premise sales calls help keep an operation foremost in potential customers' minds and can result in greater sales.

Find Out More

In nearly all cases, a foodservice operation should be a member of their local Convention and Visitors Bureau (CVB). The goal of a CVB is to promote business in a specific geographic area. CVB members typically include hotel operators, restaurant operators, and those who manage special attractions in the area, among others.

Membership in a CVB helps foodservice operators learn when large groups are coming to the area and provides opportunities to network with other hospitality industry professionals. Membership in the CVB boosts the visibility of a foodservice operation in the local community, and it also emphasizes that the operation is community-minded.

Membership in a CV is typically inexpensive and is an excellent use of a portion of a foodservice operation's overall marketing budget. To learn more about the CVB in your own area, enter "local convention and visitors' bureau" in your favorite browser and view the results.

Promotions

"Promotion" is one of the 4 Ps of marketing, but the term is also used to identify specific offers that appeal to all customers and, in many cases, provide special rewards for loyal customers. A sales **promotion** in the foodservice industry is a broad term used to identify short-term activities designed to increase business.

Key Term

Promotion (sales): A product or service offered for a limited time designed to encourage guests to make a purchase within that time.

The Purpose of Promotions

In the foodservice industry, limited-time promotions can be used to achieve several different marketing goals including:

✓ Introducing a new menu item
✓ Encouraging guests to buy an existing menu item
✓ Increasing the size of guests' purchases
✓ Increasing the frequency of guests' purchases
✓ Rewarding loyal customers

Foodservice operators can choose from many types of promotions. For example, limited-time deals or specials are promotions that offer reduced pricing on a specific product for a specific time. Coupon-based promotions may encourage customers to try an operation for the first time or to buy a selected item for a discounted price. Historically, promotional coupons were printed and distributed to customers in a variety of ways, and increasingly today digital coupons are popular with guests.

Regardless of their form, promotions are like advertising since they have a cost, and they are generally expected to yield additional sales volume to help offset the expense of offering them.

One way for foodservice operators to consider their overall sales promotion efforts is to separate them into special offerings designed for all customers and those designed for loyal customers.

Promotions for All Customers

All foodservice operations can benefit from well-designed and implemented promotional efforts directed at their target markets. Simply stated, the best promotions are designed to give customers an extra incentive to buy.

For example, consider a food truck featuring street tacos that offer a "Taco Tuesday" promotion: guests can purchase at 50% off on that day, and the incentive to visit on Tuesday is made to the truck's new and existing customers. In this example, the food truck operator likely hopes the increased volume created by the promotion will be of long-term benefit and will also encourage guests to buy tacos on days other than Tuesday.

Many promotions involve reducing prices, and they can take several forms. Time-based discounts offer reduced pricing during a specific time such as when an operation offers 50% off discounts on appetizers purchased between 4:00 p.m. and 6:00 p.m.

Flat-rate-based discounts typically offer a reduction from a guest's bill when, for example, an operation creates a coupon good for $5.00 off an entire meal. A **buy-one-get-one (BOGO)** promotion offers guests a discount that varies based on what guests purchase.

While many popular foodservice promotions are directly related to price, others are not.

Key Term

Buy-one-get-one (BOGO): A sales promotion in which an item is offered for free when another of the same item is purchased at full price.

Consider a casual service restaurant offering a special package on Saint Patrick's Day that includes a traditional Irish meal and a glass of beer for one specified price. In this example, the operation seeks to attract guests by recognizing a special event, rather than through a price reduction. Similar time-based promotions can be offered during popular holidays and/or at specific times and seasons of the year.

Promotions for Loyal Customers

A **customer loyalty program** is an effective way for foodservice operators to reward their most dedicated guests. These programs first originated in the airline industry and were quickly adopted by hotels and restaurants because of their success.

Whether they earn "Rewards," "Points," or some other designation (for example, Starbucks customers earn "Stars," and Domino's Pizza customers earn "Pie rewards"), customer loyalty programs are extremely popular with today's foodservice guests.

The best customer loyalty programs have common features. These programs:

✓ Make it easy for guests to create their accounts
✓ Provide guests with up-to-date point totals in real time
✓ Offer some items for free based on point totals earned
✓ Offer discounts on some items based on point totals earned
✓ Include basic rewards that are relatively easy to attain
✓ Include significant awards for extremely loyal guests
✓ Are tiered and allow guests to earn improved category status when additional purchases are made
✓ Make it easy for guests to redeem their points
✓ Are designed to ensure the security of guest data

When a foodservice operation creates and markets its frequent guest program, it should do so on all communication channels including the proprietary website, Facebook page, Twitter account, and all other communication channels utilized by the operation.

Key Term

Customer loyalty program: A marketing tool that recognizes and rewards customers who make purchases on a recurring basis. The programs typically award points, discounts, or other benefits that increase as the total amount of a repeat customer's purchases increase.

Also commonly known as a "Frequent guest program," "Frequent customer program," or "Frequent dining program."

Technology at Work

The best customer loyalty programs are carefully designed to create and utilize historical purchasing data to deliver targeted promotions to an operation's most loyal customers.

From a loyal customer's perspective, an operation's program must be easy to use and understand. It should also provide real-time data describing the number of points a guest has accumulated and for what those points can be used.

(Continued)

From an operator's perspective, a frequent diner program should help the operator understand what their frequent guests are buying, when they are buying, and how they are buying. When that information is known, operators are best able to tailor promotions that are valued most highly by their loyal guests.

It is essential that a foodservice operator's frequent diner program be tied directly to the operator's point of sale (POS) system. The POS is the source of the data used to update the status of guests in an operation's customer loyalty program. Effective guest loyalty management programs are simply too complex to be managed manually.

Fortunately, there are many vendors who offer software solutions designed to assist foodservice operators in the management of their frequent diner programs. To see examples of these vendors' products, enter "guest loyalty management programs for restaurants" in your favorite search engine and view the results.

What Would You Do? 8.1

"Well, it doesn't seem it would make any difference," said Martha, the assistant manager of the Oriental Orchard restaurant.

The Oriental Orchard was a Chinese food-themed restaurant that had a small dining area and did most of its business through carryout. Its "Special Dinner" was its most popular menu item, and it included an entree, fried rice, and an egg roll all sold for one price.

At the request of Ben Goh, the operation's owner, Martha was talking to Roger, the kitchen manager, about the best type of promotion to be run the same week as the just-announced opening of a major national competitor that would be located only a mile away from the Oriental Orchard.

"I understand that giving 50% off our Special Dinners like you are suggesting would be popular," said Roger, "but I think my idea of a buy-one-get-one free just makes more sense. I don't think we should get customers thinking that our prices are so low that, when they pay regular prices, they feel like they're not getting as much value."

"Well, it still seems like the same thing to me," replied Martha, "and my idea of a lower price might attract some new customers that haven't visited us before."

Assume you were Ben Goh. Do you think offering a BOGO dinner promotion during the week the new competitor opens would be the same as offering dinners for 50% off during that same week? Which promotion do you think would send the best long-term marketing message to the guests of the Oriental Orchard? Explain your answer.

Publicity

Unlike advertising, **publicity** for a foodservice operation is obtained at no cost, and it can be positive or negative. As an important part of the marketing mix, positive publicity must be carefully planned, while negative publicity must be carefully managed.

Publicity is unique among the components of an operation's marketing mix because it cannot be directly controlled. Despite that, nearly every foodservice operation can utilize publicity as an important way to improve its brand image and increase its brand awareness.

Key Term

Publicity: Free media attention for a product, service, business, or employee. It can include the use of traditional news sources such as televised news programs, magazines, and newspapers and through newer media such as podcasts, blogs, and websites.

Positive publicity is important because the increased visibility resulting from it makes potential guests more comfortable doing business with an operation, and it also builds the operation's credibility. Building an operation's name and brand awareness leads to more customers and higher sales.

Positive publicity is especially effective because most customers trust independent news sources more than they do a business's own advertising. For example, if an operation states in an ad that it is a "Great place to work," potential guests (and potential employees!) may or may not believe this claim. However, assume the operation is named in a news story as one of the top 10 "Best Companies to Work For" in an area. Hopefully, this information will be widely publicized, and guests and potential workers will trust the information and feel comfortable affiliating with an operation known for treating its employees well.

Initiating Positive Publicity

Since there is no cost, every foodservice operation would like free positive publicity; however, in many cases, it must be planned. Foodservice operators must be creative as they consider the potential of using positive publicity as a significant part of their marketing mix. The good news is that those who write and blog about foodservice-related topics are continually looking for information they feel would be important to their readers and viewers.

One way to consider the types of activities that may be of widespread interest and will reflect positively if widely publicized, is to categorize the activity as an event or an action.

Events

Special events hosted by foodservice operations can generate positive publicity. For example, a grand opening hosted by a new operation is likely to be newsworthy. Similarly, if an existing operation relocates, this event is also likely to be newsworthy.

For existing operations, a special event celebrating its 5th, 10th, 15th, or 20th anniversary is likely to be newsworthy, and this is especially so if the anniversary event offers a free menu item, discounted meals, or prizes to those who attend. For some foodservice operations, the employment of a new chef and naming of a new manager may be a newsworthy event if the announcement offers guests something special on a specific day or during a specific week.

In some cases, foodservice operators can create special events by hiring local musicians to play during a specific serving time. If reduced prices or free samples are also offered to those who attend the music event, local media interest may be significant.

Other examples of special events hosted by a foodservice operation include celebrating victories of a local sports team, welcoming parties for new business leaders in the community, and celebrating a significant event that affects the community.

To create a newsworthy special event, guests attending the event must be treated in a special way. Simply stating that an operation will hold a special event without special activities and/or promotions will not likely be considered newsworthy by local media.

Actions

Actions taken by a foodservice operation can also be newsworthy, and fund-raising efforts are an example. When a foodservice operator undertakes an important fund-raising event by donating food (or coupons for food), time, or proceeds to a popular charity, that action will likely be considered a newsworthy event. Then the operation can benefit its community while, at the same time, raising its profile and positive public image.

Showing support for local schools is another action that nearly all foodservice operators can use to generate positive publicity. For example, an operator may identify one night each month and use social media accounts to promote a fundraiser with a local school official and/or local news personality. A no-cost special soft drink or dessert could be offered for students with excellent report cards or perfect attendance. Taking a photo of each child who attends and, with their parent's permission, posting it on social media is another way to parlay this activity into positive publicity.

In nearly all cases, elementary, middle, and high school sports teams need financial support. The operation's fund-raising activities can include posting its logo and name on sports programs. Other ways to be highly visible to those attending games can also yield publicity that generates a positive reputation for the operation that exceeds costs of team sponsorship.

Partnering with a local animal shelter or animal rescue organization is another way to give back and help the community. For example, an operation might consider offering a gift card to anyone who adopts a pet during a specific time period.

Helping to feed the hungry or the homeless is another way for a foodservice operation to benefit its community and generate positive publicity. For example, an operation may designate a day, week, or month to donate to a non-profit partner. Delivering needed meals to those who work in nonprofit organizations including food banks is another positive action that demonstrates an operation's commitment to those who are less fortunate in the community.

The best positive publicity actions to take for any foodservice operation depend upon its size and location. Foodservice operations' owners and managers who think creatively, however, will be able to create events and take actions that are newsworthy and generate positive publicity.

Addressing Negative Publicity

Negative publicity is any public communication that threatens the good reputation of a foodservice operation. Negative publicity can result from several types of events that include negative publicity from:

1) A foodborne illness outbreak.

   Food poisoning or infections are a major concern for every foodservice operation and, if guests become ill after eating at a facility, the information is likely to be widely reported.

2) Negative health inspection ratings.

   In many communities, health inspection scores are posted publicly and reported widely. If an operation obtains a poor rating (low score), that information can spread quickly.

3) Inappropriate treatment of guests.

   Foodservice employees may treat guests poorly based on their race, sexual orientation, religion, or dress. When information is reported by affected guests and other diners, and/or when these events are taped on a mobile phone and widely distributed by the guest or others, widespread attention on social media and in the traditional press is likely.

4) Occurrence of a criminal event.

   Like most businesses, foodservice operations are open to all potential guests including those who intend to do harm to other guests or employees. If such an event occurs and serious injuries or death occurs, the event is likely to be widely publicized.

In any of these events (and others!), foodservice operators must rely on the rapid implementation of their **crisis management plan (CMP)**. A CMP should be developed before, not after, an event that creates negative publicity. An effective CMP is a written document with step-by-step instructions that should be referred to immediately if a crisis occurs. A CMP can be relatively brief, or extensive, depending on the needs of a specific operation. All CMPs should be written to help ensure that, if an operation is facing a crisis, it:

Key Term

Crisis management plan (CMP): Written instructions that describe how an operation will react to a crisis, including who will be involved in the response and what will be done.

1) Has designated a specific individual to address the crisis.

The individual chosen is likely to be the owner or manager, but someone else, including a trusted long-term employee, could also be selected. In most cases, an operation will only want a single person to be the spokesperson and handle all external communication, so the operation has a consistent voice and message regarding its crisis response.

2) Allows the operation to stay ahead of the issue.

During a crisis, a foodservice operation should be proactive, not reactive. It should ensure that the incident's news and updates come from the operation, and not from outside sources. Updates can be posted regularly on an operation's social media pages and shared with local media.

3) Is honest in its response.

In cases where an event is beyond the control of the foodservice operator, this can be stated. In situations where an operation made a mistake, this should be admitted. While doing so may seem to make the operation look bad, guests will appreciate the honesty. In addition, when an operation tries to deceive its customers and the public and the truth is later revealed, this can create an even more negative consequence for the operation.

4) Apologizes and explains what will be done in the future.

After an incident is truthfully explained, the operator can apologize (if applicable) and describe what will be done to correct the situation and ensure the problem does not reoccur.

5) Projects a calm and professional image.

It can be difficult to stay calm and professional in an emergency. However, doing so is especially important because being over-emotional can yield poor decision-making. Note: If the situation is managed professionally and caringly, most foodservice operations can overcome crises that are encountered.

> **Find Out More**
>
> Foodservice operations can face a variety of potential emergencies, disasters, and crises every day. Fires, floods, power outages, and similar events all call for crisis planning and pre-preparation. From a marketing perspective, however, a "crisis" is best viewed as an event that is disruptive and unexpected, and that threatens the operation's reputation.
>
> The best crisis management activities are those related to preventing a crisis from occurring. Once a crisis has occurred, however, proper response and recovery efforts are essential to help ensure an operation's continuing positive reputation.
>
> The Internet is a good source of information about the ways foodservice operators can prepare for and respond to a reputational crisis. To review this information, enter "foodservice crisis management planning" in your favorite browser and view the results.

Public Relations

Public relations (PR) include all activities designed to ensure that a foodservice operation has a positive public image. PR is not the same as publicity. While publicity consists of public information about an operation that is typically beyond the operation's direct control, PR allows an operation to contribute to and even shape the communication that is shared with the media.

Foodservice operations use PR activities to inform potential guests that they are good community citizens. Examples include hosting charity events, contributing cash or in-kind services to worthy causes, and volunteering staff time for worthy causes.

> **Key Term**
>
> **Public relations (PR):** The various ways a foodservice operation communicates with the media and the public to build mutually beneficial relationships; also commonly referred to as "PR."

Effective PR activities can shape and maintain an operation's positive reputation that is formed by what it does, what it says, and what others say about it. Potential guests want to do business with operations that enjoy positive reputations while, in contrast, they often want to avoid visiting facilities with a poor reputation.

Media Affiliations and Successful Public Relations

Positive affiliations with members of the media are essential to successful PR efforts because they can help shape and communicate a positive reputation for an operation. In many cases, these are the same persons who can help overcome any unwarranted negative publicity encountered by the operation.

In the last decade, communicators have gone from a single communication tool (**press releases**) to dozens of social media options to reach journalists.

Key Term

Press release: A communication tool used to objectively announce something newsworthy. Its purpose is to obtain media coverage and be noticed by an operation's target market.

Members of the media are always looking for stories to interest their audiences. For example, assume a new food trend is becoming popular. If a specific reporter (or blogger) knows a foodservice operator who can provide timely and valuable information about the topic, this media member is more likely to use the operator as a source rather than someone else. Additionally, if the operator is consistently honest and informative, the relationship between the reporter and operator will likely be positive.

Media professionals can serve as advisors to operators and let them know how they can best communicate newsworthy events occurring in their operations. Examples may include hiring a new chef, introducing new menu items that follow national trends, and other topics of local interest.

Foodservice operators should reach out to important local media professionals as part of their marketing mix activities. Inviting them to their operations and communicating with them regularly is important. When professionally initiated and nurtured by both parties, the relationship between a foodservice operator and a media professional can benefit both parties.

Community Relations and Successful Public Relations

Community relations include activities a foodservice operation undertakes to maintain its position as a good neighbor and productive member of its community.

Key Term

Community relations: The work an operation does to establish and maintain goodwill in its community.

A foodservice operation's community relations efforts including acting ethically and contributing to civic priorities are good for business. This, in turn, can foster goodwill among those in the community who are the operations' potential customers.

Foodservice operations who donate products, money, or time to worthy local causes can help their communities thrive, and operators who generate positive word-of-mouth communications resulting from these efforts can boost an operation's credibility. This, in turn, creates a personal connection with target markets as community relations efforts verify that the operator is compassionate, civic-minded, and interested in helping with local community improvement activities.

Having a positive reputation in the community also makes it easier to find, and retain, high-quality staff members. When an operation effectively employs its community relations efforts to enhance its reputation in the community, workers will view the operation as dependable, trustworthy, caring, and honest. These are characteristics all workers, especially younger workers, increasingly look for in their employers.

Employee Relations and Successful Public Relations

Effective public relations is all about an organization's reputation in its community. From that perspective, every employee in a foodservice operation has an opportunity to present his employer in a positive light—or in a negative one—when he is away from work.

Like other employees, their places of employment are popular topics with foodservice workers at all levels. Discussions about their supervisors and experiences at work are routinely shared with family, friends, and other foodservice workers. When the experiences shared reveal, an employer is fair, honest, and kind, that information will be communicated. Similarly, however, if an employer is perceived to be unfair, harsh with its workers, and interested only in themselves, that information will be shared as well.

When employees are treated fairly, they feel good about coming to work. In contrast, if they are treated unfairly, they will likely dread coming to work. Treating all employees with respect and showing genuine appreciation for their efforts will result in the employees treating their customers with respect and appreciation. Experienced business professionals know most employees want to feel they are doing more than simply contributing to their employer's bottom line. When workers see their employers as active and positive members of the community, they will have significant pride in their affiliation with that employer, and this in turn, will result in better employee performance and reduced employee turnover.

All foodservice operators choose from several activities as they seek to best promote their operations. Figure 8.3 provides a summary of the five marketing mix activities available to all foodservice operators.

As they plan their varied promotion efforts, foodservice operators increasingly recognize the importance of using web-based tools to communicate with their guests. The single most important communication tool for all foodservice operations is their own website. For that reason, effective web-based marketing on an operation's own proprietary website will be the sole topic of the next chapter.

Activity	Goal
Advertising	Delivering the marketing message in a cost-effective manner
Personal selling	Increasing customer counts, operational visibility, and check averages through face-to-face interactions with customers
Promotions	Providing all guests and especially loyal customers with special buying opportunities or rewards
Publicity	Increasing attention from the media at no cost to the operation
Public relations	Increasing name recognition and awareness of an operation in its own community

Figure 8.3 Marketing Mix Activities Summary

What Would You Do? 8.2

"Who was that on the phone?" asked Margarida.

"Another reporter. This one was from the local newspaper," replied André.

Margarida and her husband André were the owners of a fast-casual franchise operation that featured South American cuisine. Business was good, and they had a steady flow of customers. That was, until last week when a foodservice operation within the same chain was found to be responsible for a foodborne illness outbreak caused by improperly handled raw chicken.

The operation responsible for the outbreak was in an adjacent state, but the publicity from the outbreak resulted in several calls from local television and newspaper reporters. Their questions focused on whether Margarita and Andre felt their customers could still safely eat in their restaurant.

"It's not fair," said André, "We always get excellent reports from the health department, and we've never had any sanitation-related problems of any kind in the past."

"Is that what you told the reporter?" asked Margarida.

Assume you were Margarida or André. How would you respond to a reporter who asked if it was safe for your customers to eat in your operation? Do you think their excellent sanitation reports would be good information to share with the reporter? What other information could be shared?

Key Terms

Advertising	Advertising campaign	Frequency (advertising)
Grand opening	Reach (advertising)	Personal selling

Up-selling	Customer loyalty	Public relations (PR)
Sales call	program	Press release
Promotion (sales)	Publicity	Community relations
Buy-one-get-one (BOGO)	Crisis management	
	plan (CMP)	

Operator's 10-Point Tactics for Success Checklist

Evaluate your need for, and the current status of, each of the following operational tactics. For those tactics you think are important, but not yet in place, develop an action plan for its implementation including who will be responsible for the tactic's completion and the target date by which it should be completed.

				If Not Done	
Tactic	Don't Agree (Not Done)	Agree (Done)	Agree (Not Done)	Who Is Responsible?	Target Completion Date
1) Operator recognizes the key role of paid advertising in the overall marketing message mix.	____	____	____		
2) Operator has reviewed available communication channels and understands the impact of reach and frequency in each of them.	____	____	____		
3) Operator understands the importance of developing and implementing an effective on-premise up-selling program for service staff.	____	____	____		
4) Operator reserves some time each week for off-premise personal selling activities.	____	____	____		
5) Operator recognizes the implementation of well-timed promotions that target all guests as an important part of their operation's marketing mix.	____	____	____		

(Continued)

Tactic	Don't Agree (Not Done)	Agree (Done)	Agree (Not Done)	If Not Done	
				Who Is Responsible?	Target Completion Date
6) Operator has developed, implemented, and properly maintains a guest loyalty program appropriate for their own operation.	____	____	____		
7) Operator recognizes the value of free publicity in the overall marketing mix.	____	____	____		
8) Operator's marketing plan includes details for implementing specific positive publicity events and activities.	____	____	____		
9) Operator has developed a crisis management program to be used in the event the operation encounters significant negative publicity.	____	____	____		
10) Operator understands the importance of public relations efforts undertaken with members of the media, the community, and its own employees.	____	____	____		

9

Web-Based Marketing on Proprietary Sites

What You Will Learn

1) How to Choose a Primary Domain Name
2) How to Choose a Web Host
3) How to Create a Proprietary Website
4) The Importance of Search Engine Optimization
5) The Importance of a Social Media Presence

Operator's Brief

In this chapter, you will learn that in today's foodservice environment, an operation's proprietary website is its single most important marketing tool. The process of developing an effective proprietary website begins with identifying an operation's primary domain name, and this must be done carefully.

In most cases, a foodservice operation will not buy its primary domain name outright. Instead, it can be rented for a fixed time. In many cases, the rental fee for a domain name will be waived when an operator selects a web host; the organization that will maintain and deliver the content to be displayed on an operator's proprietary website.

The design and creation of an effective proprietary website is not typically undertaken solely by a foodservice operator. Instead, it will be created in conjunction with professionals who specialize in website development. Sometimes a foodservice operator's website development will be undertaken in conjunction with an operation's web host, but this does not always occur. Regardless of who helps, operators must develop their proprietary websites with a focus on design, content, and the ability to track website traffic. An effective website will include some mandatory and, in many cases, some optional features.

(Continued)

The optimization of search engine results is a critical aspect of website design. The selection of keywords and phrases will have a direct impact on an operation's placement on a search engine results page (SERP), so these terms must be chosen carefully.

In addition to the content placed on their proprietary websites, operators increasingly place information on social media sites. The popularity of such sites continues to grow, and these sites should be an important part of every operation's overall Internet-based marketing efforts.

CHAPTER OUTLINE

Choosing the Primary Domain Name
Choosing the Website Host
The Proprietary Website
    Website Design
    Website Content
    Website Traffic Tracking
Search Engine Optimization
    How Search Engines Work
    Choosing Keywords and Phrases
Social Media Sites
    The Importance of Social Media
    Posting Content on Social Media Sites

Choosing the Primary Domain Name

To establish a web presence, every foodservice operation must begin by selecting its **domain name**. An operation's domain name is second in marketing importance only to choosing the operation's actual name (see Chapter 3).

Foodservice operators need not be computer experts to understand the importance of their domain names. Essentially, computers rely on a language when they communicate with each other, and each computer device is identified with its own unique **IP address**. For example, a foodservice operation's actual IP address might look like the following:

192.168. 123.132.
Or
3ffe:1900:4545:3:200:f8ff:fe21:67cf

Key Term

Domain name: The unique name that appears after www. in web addresses and after the @ sign in e-mail addresses.

Key Term

IP address: A unique address that identifies a device on the Internet or a local network. IP stands for "Internet Protocol," which is the set of rules governing the format of data sent via the Internet.

To navigate easily around the Internet, typing in a long IP address to connect to a website is not realistic for online users. Actual IP addresses are complex and not easily remembered. This is the reason domain names were created: to convert IP addresses into something more easily remembered. Therefore, an operation's domain name can be considered a "nickname" for its IP address. Put another way, domain names provide a human-readable "address" for a foodservice operation. The address chosen must be unique and cannot be an address already used by another person or entity.

Selecting a domain name to help market an operation must be done thoughtfully, and there are several key steps and tips. These include:

1) Choosing an **extension**

    A foodservice operation's chosen domain name will be followed by an extension (for example, followed by .com., .net., .org., or others.)

Key Term

Extension (domain name): The combination of characters following the period in a web address.

Today, there are hundreds of domain name extensions available in addition to ".com," which is the original and most common extension. Despite the fact that both ".rest," and ".restaurant" exist and are used by some operations, in most cases, it is best if an operation uses ".com" as its extension. The reason: most guests searching for a restaurant will be familiar with ".com" and may not be familiar with other extensions. In fact, many users, especially those who are not very tech-savvy, automatically type ".com" at the end of every domain name without even thinking about it.

2) Keeping the domain name short

    Some foodservice operators want to put as many words and as much information in their domain name as possible. For example, a New York-style pizza operation named "Tony's" might want to use a domain name such as:

"Tony's, home of the best New York-style pizza in the city.com"

This domain name is descriptive and includes the operation's name and a statement about the quality of the pizza sold by the operation. However, the name is too long and would not be easily remembered by most potential guests. Note also that entering the operation's web address would require the use of an apostrophe between the "y" and "s" in the name "Tony's," a comma after the word "Tony's," and a dash (hyphen) between New York and style. Most potential users would not know about these requirements to find the website address.

As well, Internet marketing experts typically recommend that a domain name consist of 15 or fewer characters because when a domain name is too long, users increasingly make mistakes (typos) when entering the longer name.

In this example, a better domain name choice might be:

Tonyspizza.com

Or, if that domain name is taken,

Trytonyspizza

Note that both domains are short, simple, and would likely be remembered by most guests.

3) Keeping the domain name easy to pronounce and spell

As noted, when selecting the name of a foodservice operation (see Chapter 3), the domain name must be easy to pronounce and spell. Operators selecting their domain names should avoid foreign words and language terms unfamiliar to many Internet users, and another suggestion is to avoid all "hard to spell" words.

4) Avoiding hyphens and double letters

Some operators may need to create their own unique domain main name when their first choice has been selected by another operation. For example, if a donut shop operator named Bob finds that the domain "Bobsdonuts.com" has already been taken, the operator may be tempted to select a domain name of "Bobs-donuts.com."

This domain name would likely be available, but it is also likely that most guests looking for "Bob's Donuts" in this example would go to the original domain name site, rather than the operator's hyphenated site (hyphenated domains are also prone to typos). Similarly, domain names with double letters are prone to user entry typos. In most cases, avoiding double letters make a domain name easy to remember, easy to type and, therefore, more brandable.

5) Researching the proposed name

Before finalizing a domain name, operators must determine if there is already a business using the same name. If so, the operation will not be allowed to use the name. When researching domain name availability, operators should perform a Google search and check for the name's availability on top social media websites including Twitter, Facebook, Instagram, and others.

Key Term

Registration (domain name): The act of reserving a domain name on the Internet for a fixed period of time, usually one year. Registration is necessary because domain names cannot typically be purchased permanently.

After an operator has selected an available domain name, the next step in using it is its **registration**.

Technology at Work

In most cases, the use of high-quality domain names is not free. A domain name must be rented or, in just a few cases, purchased. Typical rental fees for a domain name range from $2 to $20 per month.

There are many organizations that assist operators in searching the Internet to determine whether a proposed domain name is already used, or whether it is available for use. They can also provide suggestions about alternative domain names that are memorable and available.

To see how these search sites work and to learn how to use them, enter "search for domain names" in your favorite search engine and view the results.

Domain name registration is required for a website, for e-mail, or for other web services. An operation's domain name can be used if the rental payments continue for it. In most cases, a **web host** will waive rental fees for a domain name if an operator agrees to select that organization to host its website.

Key Term

Web host: An organization that provides space for holding and viewing the contents of a website. All websites and e-mail systems must be hosted to connect an operator's proprietary website content to the rest of the Internet.

Technology at Work

Domain name registration gives an operation a recognized identity, and many organizations can register an operation's chosen domain name. After a domain name is registered, the information about its owner is publicly available.

Domain names are registered by registrars accredited with the Internet Corporation for Assigned Names and Numbers (ICANN); the nonprofit organization responsible for coordinating all numerical spaces on the Internet.

To see how these registration organizations work and to learn about the services they offer in addition to domain name registration, enter "how to register domain names" in your favorite search engine and view the results.

Choosing the Website Host

The web host selected by a foodservice operator will become an important marketing partner, and Figure 9.1 illustrates why this is true. A web host not only receives content from a foodservice operator but also distributes that information

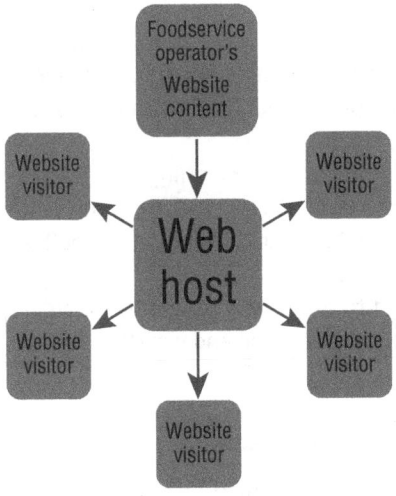

Figure 9.1 Web Content Distribution

to those visiting the operator's website. The ability to effectively manage both key tasks is the mark of a high-quality web hosting partner.

Web hosts play a key role in disseminating information about a foodservice operation because they store information about an operation in web servers that are connected to networks. When a visitor types in a web address, the Internet transmits the request to the web host. The host then (i) retrieves the operation's files and (ii) sends the information back to the visitor's computer so the site can be viewed.

There are important factors to be considered when selecting a web host. These include:

1) Speed

The speed at which an operation's website loads is important for making it easy to use. Today's foodservice guests will have little patience with a slow-loading website. Speed of download is especially important if a foodservice operation includes a video on the site because these files take longer to download than information presented in regular or **PDF** text.

Key Term

PDF: Short for "Portable Document Format." A document format that allows documents to be shown clearly regardless of the software, hardware, or operating system utilized by the viewer.

2) Reliability

The typical foodservice operator does not have detailed knowledge about the specific servers and networks used by a web host. However, the chosen web host should be reputable and able to guarantee consistently high levels of speed and reliability.

3) Security

The security levels offered by a web host are extremely important, especially when a website will process guest payments or send personal information to the website. (Consider, for example, when guests participating in a guest loyalty program enter personal data on the website.)

4) Support

The operation's website is increasingly an important marketing tool, and problems such as an inability to download content, blank or missing pages, or related issues must be resolved quickly. The best web hosts have 24/7 support that is easily accessible and that can quickly lead to the resolution of content display issues. As an operation's website increases in content and functionality, the importance of high-quality support increases as well. The best web hosts offer a choice of support contact methods including live chat, telephone, and e-mail.

5) Data Management Tools

The best web hosting sites provide operators with data management tools some of which are at no cost (they are included as part of the web host's monthly fee) and in other cases, they are available for purchase.

Essential tools that should be provided by a web host include:

- A Content Management System (CMS) program used to easily make changes to website content (see Chapter 7)
- A Customer Relationship Manager (CRM) program to manage customer loyalty programs (see Chapter 8)
- A website visitor tracking system to gain important information about a site's users
- E-mail services (optional) but often included at no charge

Many web hosts offer foodservice operators a package that includes hosting and domain name registration. Some operators purchase their hosting and domain name registration from the same company. One possible concern: if the operator wants to use another web host in the future it can be difficult to separate the domain name from the original web host. Some Internet-marketing experts recommend securing the domain name separately so web hosts could be easily changed if an operator desires to do so.

Technology at Work

GoDaddy, Amazon Web Services, and Google Cloud* are currently the largest web hosts. These may fit the needs of a specific foodservice operator and should be considered as potential web hosts. However, there are likely other web hosts that might better address the operator's needs.

The best web-hosting decisions are made when operators know what their website should be able to do. For example, will guests visiting the website be able to do online ordering and payment, or will they simply be viewing a menu in a print format? Similarly, will guests be able to make a reservation on the website or join and review guest loyalty program points?

The features and functions to be offered on a foodservice operation's website most often help determine the best web host. To see a list of available foodservice-oriented web hosts that could be considered before making a final selection, enter "best web hosts for restaurants" in your favorite browser and view the results.*

*Ranking as of 7-12-2022: https://www.hostingadvice.com/how-to/largest-web-hosting-companies/

The Proprietary Website

A foodservice operator's **proprietary website** is one in which the operator has full control over the website's content.

Regardless of their service style or size, all foodservice operators should create and manage a professionally developed proprietary website that will be found at

> *www.registered operation domain name.*
> *chosen ext*

Key Term

Proprietary website: A website in which the foodservice operator controls all the website's content and can readily make changes to it.

The marketing-related goals of a foodservice operator's propriety website are easy to identify. These include:

1) They show the operation's NAP

   One function of a website is to show an operation's name, address, and phone number (NAP). This is important whether an operation is independent or part of a chain. If it is an independent, and particularly a start-up, potential guests must know the name of the business, where it is located, and how to contact it. If a foodservice operation is part of a local or national chain, a proprietary website identifies the specific operation's location.

   Previously, foodservice operations disseminated their NAP information in the Yellow Pages, a printed directory distributed by telephone companies. Today's customers search for NAP information on their computers, tablets, and smartphones, and a proprietary website enables them to learn about a business and what it offers anytime and at any place they choose to do so.

2) Attracting customers

   Successful foodservice operations maintain their current customer bases and expand them with new customers. Reaching thousands of potential new customers using only traditional marketing methods such as direct mail, print advertising, or radio can be time-consuming and expensive.

   A strong online web presence enables operators to reach more people locally while paying a relatively modest amount for exposure. In most cases, potential guests would rather look online than call a business to learn basic information such as its operating hours or location. Operations without websites miss out on all potential customers who do not want to make a phone call to ask about offered products and services.

3) Building credibility

   Many customers using the Internet to research a business have likely viewed thousands of different web pages, and they can easily differentiate between a high-quality website and one that is poorly designed. The quality of an operation's website communicates much information to those who surf the web.

When a website appears professionally designed and is easily navigable and visually exciting, it enhances the foodservice operation's image. Guests viewing the website often believe a professionally prepared website suggests the operation is run professionally, and that it is established and successful. This straightforward communication of business stature builds an operation's reputation and enhances its credibility. It is simply a fact that most of today's guests would not consider a foodservice operation without any type of Internet presence.

4) Growing the business

In many cases, a potential customer will not know the domain name of an operation they desire to visit, and the operation's proprietary website then becomes particularly important. To illustrate, assume a family visiting a new city wants to visit an Asian-style restaurant offering Thai cuisine. By performing a **Google search**, this family would be directed to one or more proprietary websites created by various foodservice operations in the city that offered Thai cuisine.

Key Term

Google search: Also known simply as "Google;" this is a method of finding information on the Internet. Note: Google conducts over 92% of all Internet searches and is the most visited website in the world.

After reviewing the information on alternative websites, the family decides which operation to visit. Their decision may be based on price, location or, in many cases, on the professional appearance of the website itself. If a Thai food operator in the city did not have a proprietary website, the travelling family would not choose them.

5) Gaining a competitive advantage

Some foodservice operators believe that if they have a Facebook page or are otherwise utilizing social media, a proprietary website is not of importance, but they are mistaken. The importance of a social media presence is addressed later in this chapter, and you will learn that social media platforms are restricted in terms of design, process, and technology.

A foodservice operator desiring to provide specific information and services to guests typically learns that a proprietary website assists them in doing so while gaining a competitive edge. The reason: a website not only attracts new guests but it can also be used to better serve existing guests. The ability to update operating information such as new menu items and specials on a 24/7 basis improves current guest communication, helps build customer loyalty, and enhances an operation's brand image.

There are several ways to examine key components of an operation's proprietary website, and one way is to assess three important website characteristics:
1) Design
2) Content
3) Traffic Tracking

Website Design

Few foodservice operators have the expertise and time required to design their own proprietary websites. In some cases, website design assistance is offered by a web host offering basic website templates. Alternatively, operators use the services of a professional website designer or design organization.

Professional web designers use several tactics to create dynamic and effective websites, and they pay particular attention to three key areas:

✓ Appearance
✓ Reading pattern
✓ Navigation

Appearance

The overall appearance of a proprietary website is determined primarily by color, typestyle, and imagery. The basic colors used in website development should fit with an operation's service style and, in most cases, should be limited to five colors or less. The colors used on the website should be complementary. If colors are used to identify an operation's brand, either in its uniforms, logo, or signage, these are the website colors that will best reinforce the operation's brand image.

The typestyle used to create the website should be of a size that is easily read and especially so when an operation's menu will be posted online. In most cases, professional web designers recommend that all typestyles used should be easy to read and contain a maximum of three different fonts (see Chapter 7).

The imagery used in a professional website can vary from high-resolution photographs to full videos. The imagery used in all forms of graphics should be expressive and capture the personality of the company. Since a website visitor's initial impressions are most often formed visually (not by reading printed content), the imagery used on the website is of extreme importance.

Reading Pattern

The most common way for visitors to scan text on a website is by using an **F-shaped scanning pattern**.

Key Term

F-shaped scanning pattern: A text scanning pattern characterized by fixations concentrated at the top and the left side of a page. Website viewers typically first read in a horizontal movement across the upper part of the content area. Note: This initially forms the F's top bar, and viewers then scan down. The "F" in F-shaped scanning pattern is short for "Fast," and the method is also commonly referred to as the "F-based reading pattern."

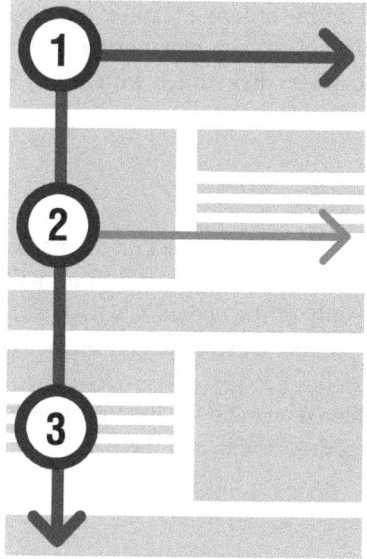

Figure 9.2 F-shaped Scanning Pattern

Figure 9.2 is a visual representation of the F-shaped scanning pattern utilized when viewing information displayed on a smartphone.

Professional website designers should be very familiar with the F-shaped scanning pattern and present website content in a way that is complementary to the reading style of the website's users. It is especially important to consider the F-shaped scanning pattern when designing websites to be mobile-device friendly since these are the types of devices increasingly used to view websites.

Navigation

Navigation refers to the process of a user finding information of importance to them when visiting a website. When websites are easy to navigate, the **user experience** is positive. Then viewers can readily find important information or accomplish a desired task (such as placing an order or making a reservation) with relative ease.

Proper website navigation contributes to a positive user experience by presenting an enjoyable-looking layout that is intuitive to use and which allows users to find the information they want as quickly as possible. When navigation on a website is poor, users will find the website confusing and will often give up their search for information and look elsewhere for what they need.

Key Term

Navigation (website): The process of moving from one content section of a website to another section.

Key Term

User experience: The feelings resulting from a user's interactions with a digital product.

Most professional web designers working with restaurants recommend that menus be split into lunch and dinner if they are different. In addition, separate wine beer and spirits menus keep the food menus from being overwhelming and "too busy." Design features that include a "call-to-action" button that encourages guests to book a table, order online, or buy a gift card, must also be easy to use.

If an operation has multiple locations, the proprietary website should include links specific to the location the customer is searching for. Different directions and phone numbers, menus (if they vary), and online ordering features must be clearly identified. Then, for example, guests do not place a pickup order with "Tony's Pizza South," when they intended to place their order with "Tony's Pizza North."

Website Content

The actual content on a proprietary website can vary widely based on the needs and interests of the operation and its ownership. In some cases, a website will offer extremely limited information consisting only of the operation's NAP, and perhaps a PDF of its menu(s). In other cases, a proprietary website may include a large number of features and functions.

Some content and features might be considered mandatory because guests expect every foodservice operation's website to address them, and these include:

NAP
Operating hours
Driving directions
Link to Google maps (to show an operation's map location)
Menu items offered
Menu item prices
Payment forms accepted
Links to social media sites

Other content areas may be included on the website if an operator determines them to be of value. Examples of optional content areas can include:

Online Ordering

While online ordering is not yet considered a mandatory content area for all operations, many foodservice operators believe this feature is or soon will be required on the website. While the future of takeout and delivery ordering by guests is still being written, operators learned during the COVID-19 pandemic that they absolutely need a way for guests to place orders without being on-site.

The superiority of online ordering versus telephone-only ordering is continually proven because an online digital record of the order reduces guest misunderstandings, confusion or errors made in the kitchen, and payment disputes. An online ordering function coupled with online payments can help manage malicious or mischievous order placers with no intention of picking up orders or who are placing unwanted orders for others.

Special Order Pickup Instructions
An increasing number of guests order menu items for pickup in many foodservice operations. An operation accepting pickup orders can make it easy for guests to get them by providing special instructions about where to park, where to go, and who to see when takeaway orders are picked up. This feature is especially important during busy service times when arriving guests may not know special arrangements have been made for order pickups.

Online Table Reservations
This feature allows website viewers to make a reservation including date, time, and number in the party. When reservations are easy to make site visitors have no need to go to another website, and the best of these reservation systems even allow website visitors to choose their preferred table in the dining room.

Gift Card Purchasing
Gift cards are extremely popular both for the person purchasing them for their own use and for giving as gifts. When an operation offers gift card purchasing online, they make it easy for a visitor to buy something they will likely need and want now or in the future. Note: When promotional discounts are offered (for example, a $50 gift card can be purchased for $45), online gift card sales can be significant.

Customer Loyalty Program Enrollment
Operators with a customer loyalty program can offer website viewers the opportunity to enroll in the program online. Providing contact information such as an e-mail address and/or telephone number increases the size of the loyalty program's membership, and it is easier for the operator to send future communications to its most loyal guests.

Operators selecting this option should recognize concerns about data privacy issues and fear of fraud. As well as changing communication preferences. For example, some guests may consider getting text messages to be somewhat invasive. Note: Where younger generations are more accepting of receiving text communication, many in older generations are not.

Even though text messages are a direct way to make sure a customer sees an operator's messages, some guests are more likely to "opt out" of receiving those messages in the future since unsubscribing to text messages consists of just responding "No" to future messages, as opposed to unsubscribing from an e-mail, which is a longer several step process that many guests will not take the time to do.

On- and Off-site Catering Information

For operations offering on-site banquets for large groups or who have mobile operations providing off-site catering services, information about these services should be posted online. Visitors to a foodservice operation's proprietary website should never have to wonder whether they can make a large pickup order or host a large event at the operation.

An Online Promotion Feature

Online promotions (see Chapter 4) are offered only to guests who view the operation's website. These promotions make the website's viewer feel "extra special," and promotions can be changed regularly to encourage visitors to return often to the website. This, in turn, further establishes the brand's online identity and enhances customer loyalty.

On-site Video

The adage that "A picture is worth a 1000 words" applies to foodservice websites, and high-resolution photos of menu items and interiors are generally essential. However, short, on-site videos of real guests enjoying real food make an especially powerful advertising statement. It also projects a level of professionalism in the foodservice operation that is not projected by photos alone. As previously noted, these videos need to be short to be easily loaded for viewing on smart devices.

Guest Reviews

This optional feature is listed last, but not because it is last in importance! In fact, as most operators agree, guest reviews are becoming the single most important factor considered by potential customers when they select a foodservice operation for the first time.

The importance of encouraging and displaying positive user-generated reviews on a website and social media and the need to respond to any negative reviews are essential to an operation's success, and they are the topics of Chapter 11.

What Would You Do? 9.1

"I don't think we have a choice," said Joyce. "If we don't go for it, we're going to continue to lose out." Joyce, the manager of Chicken Bones restaurant, was talking to Teddy Flood, the restaurant's owner.

Chicken Bones was famous for its chicken wings served with a variety of made-on-site sauces. Business was good and increasingly guests were placing their orders online for pickup or delivery. Many of the delivery orders were placed on third-party delivery sites and apps such as Door Dash and Grubhub. In most cases, these orders were paid for at the time the order was placed, and guests liked the convenience of being able to do so.

Customers who picked up their orders at the restaurant paid for them at the time of pickup because the Chicken Bones website did not provide an online payment option at the time orders were placed directly with the restaurant. When it was very busy, the wait times for order pickup were extended because of the time it took to process individual payments for the pickup orders.

Joyce was recommending to Mr. Flood that they add an online payment feature to their operation's proprietary website. This feature would allow pickup order guests to pay for their items at the time they placed their online orders.

"How does that work?" asks Mr. Flood. "I mean, how is the sale automatically recorded in the POS system? And what about payment card security?"

Assume you were Joyce. Do you think the questions asked by Mr. Flood are valid? What additional issues should Joyce address as they consider adding this new feature to the operation's website?

Website Traffic Tracking

Traffic tracking can tell operators much about their proprietary website and those who have visited it. Traffic tracking is also important because it provides valuable information to modify and improve an operation's proprietary website.

Every operator with a proprietary website can benefit from tracking the site's visitor data. Google Analytics is the best tool for website traffic monitoring. Many of its services are free and easy to use. The three most important metrics that Google Analytics provides are:

✓ Average time a user spends on a website
✓ Sources that direct users to the website
✓ Interests of the website's viewers

Google Analytics is a great way to track these metrics. However, to take full advantage of Google Analytics, a website must be developed with a design that embeds the computer coding necessary to track activity on each page.

Key Term

Traffic tracking (website): Information that tells website operators how many site visitors it has had, how the site was found, how long the visitors browsed different pages on the website, and how frequently the visitors were converted into customers.

Web traffic is carefully measured to help evaluate websites and the individual pages or sections within them. While there are several website metrics of importance to operators, most marketing professionals suggest that, in addition to other metrics, the **click-through rates (CTR)** and the **bounce rates** of a website should be regularly analyzed.

Among additional website traffic metrics that may be monitored are:

✓ The total number of visitors to the website
✓ The average amount of time each visitor stays on the site
✓ The average length of time each page is viewed
✓ The most popular website pages
✓ The least viewed website pages
✓ The source of website visitors
✓ The geographic locations of visitors that provided the most online orders

Key Term

Click-through rate (CTR): The number of clicks an ad or promotion receives divided by the number of times the ad is viewed. The formula used to calculate CTR is:

$$\frac{\text{Number of Clicks}}{\text{Number of Views}} = \text{CTR}.$$

For example, if an operator's promotional ad on a website had 5 clicks and 100 views, the CTR would be 5/100 = 5%

Key Term

Bounce rate: The percentage of visitors who view only one page before exiting a website.

Find Out More

While Google Analytics is the most well-known traffic tracking tool available to foodservice operators, it is not the only online assistance available. There are other organizations that offer traffic-tracking analysis.

The Google Analytics navigation menu has more than 125 different reports an operator could choose from to assist in analyzing their website. Google Analytics gathers data on more than 250 different web traffic metrics. For some operators, using Google Analytics without help can be overwhelming, and they desire a simpler tool. Such tools do exist and some of them use data from Google Analytics.

To see some foodservice-oriented web traffic tools and services that could be of great value to a foodservice operation of any size, enter "web marketing vitals for restaurants" in your favorite browser and view the results.

Search Engine Optimization

Every foodservice operator would like a website that is vibrant, easy to use, and effective. To be effective, it must first be seen. For it to be seen, it must be easily found on the Internet. In some cases, foodservice operators either do not understand or discount the importance of search engine results when their websites are created.

A **search engine results page (SERP)** is the list of results that a search engine returns in response to a user's specific word or phrase query.

It is extremely important that an operation be near the top of a SERP when it is viewed, and proper website design includes paying careful attention to **search engine optimization (SEO)** strategies and principles.

Marketing professionals with experience in SEO strategy assist operators by ensuring their websites' placement on SERPs is as good as it can possibly be. This is critically important because, no matter how professionally it has been developed, a website can only serve as a positive marketing tool when it is first seen by users!

Key Term

Search engine results page (SERP): The webpage a search engine such as Google, Bing, or Yahoo shows a user when the user types in a search query.

Key Term

Search engine optimization (SEO): The process of improving and modifying an operation's website with the purpose of increasing its visibility when users search for products or services related to the operation.

How Search Engines Work

Foodservice operators need not be experts in SEO to recognize its importance. However, a basic understanding of how search engines return their results in response to a user query is important to high-quality website development.

Essentially, search engines such as Google are continuously scanning webpage content placed on the Internet. The engines move from site to site, collect information about those pages, and develop it in an index. The index can be visualized as a giant library where the librarian (the search engine in this scenario) can immediately pull up a book (a webpage) that includes exactly the information a library visitor (the computer user) requests.

Next, algorithms analyze pages in the index by considering hundreds of signals and pieces of information provided on web pages. This is done to determine the order the pages should appear in the SERP for a given computer user's query.

Returning to our library analogy, it is as if the librarian (the search engine) has read every single book in the library and can immediately tell a user which ones have the proper answers to their questions. Unlike advertisements, search engines will not accept money to have an operation's web page appear higher on a SERP than do other websites. SERPs can, however, be directly influenced by the **keywords** and phrases operators place on their proprietary webpages.

Key Term

Keywords: The words and phrases that users type into search engines to find information about a particular topic. Also known as "SEO keywords," "Keyphrases," or "Search queries."

Every SERP produced is unique, even for search queries performed on the same search engine using the same keywords or phrases in their search queries. This is because search engines customize the experience for their users by presenting their SERPs results based on additional factors beyond the user's search terms. These additional factors can include the user's physical location, browsing history, and social media settings. Therefore, operators must recognize that while two SERPs may appear identical and may contain many of the same results, there will most often be subtle differences that can be important.

Choosing Keywords and Phrases

As explained above, keywords are any term used in a search, and they can be a single word or a long phrase. For example, the word "Pizza" is a keyword, but so is the multi-word phrase "Pizza near me now." As shown in Figure 9.3, anything typed or spoken into a search engine's query bar can be considered a keyword.

A detailed examination of how professional website designers collaborate with foodservice operators to optimize SERPs for their unique proprietary websites is beyond the scope of this chapter. There are, however, some principles that all operators can utilize as they place keywords and phrases into the content displayed on their proprietary websites.

To select keywords and phrases that assist in SEO, operators should:

1) Think like customers

Search engines are designed to be user friendly, and key words and phrases anticipate normal language and queries. When determining the best keywords and phrases, operators should ask themselves; "If I wanted to find my operation what would I type into Google?" Talking to other foodservice professionals, staff, and current customers can help generate opinions about phrases that would likely be used to search for a specific operation.

Figure 9.3 Google Keyword Search

2) Consider search volume

In some cases it may be possible for a foodservice operation to be listed at the very top of a SERP, but it is not a SERP that many customers ask to see. In this case ranking first for a search word that no one ever uses is an ineffective use of a keyword. Operators should avoid the extensive use of unfamiliar terms and those that are hard to spell.

3) Use available keyword research tools

There are a number of free SEO tools that can help operators gain insight into their website traffic. By using Google Analytics, Keyword Planner, and Google Search Console, for example, operators can gather data on keyword volume and current search trends. The use of keyword research tools can provide invaluable information on how to choose keywords.

4) Study competitors' websites

Nearly all foodservice operations have direct competitors. It will be worthwhile to look at these competitors' proprietary websites to see the keywords and phrases they are using because doing so might help an operator to include terms or descriptions they might have otherwise forgotten.

5) Select keywords that are also suitable for social media use

When an operation has a presence on social media (as most should!) utilizing the same keywords on the proprietary website as on the operation's Facebook, Instagram, or Twitter accounts helps optimize search engine results on these sites.

Find Out More

Search engine optimization (SEO) and the selection of the best keywords to optimize a foodservice operation's placement on a SERP are both an art and a science, and SEO strategies are ever-evolving.

SERPs are the web pages displayed to users when they search for something online using a search engine such as Google. The user enters a search query, often including specific terms and phrases referred to as "keywords." The process of creating SERPs is constantly changing due to on-going experiments conducted by Google, Microsoft, Bing, Yahoo, and other search engine providers to offer users a more intuitive and responsive experience.

As search engine technology changes so too will the ways they generate their SERPs. To follow changing SEO technology and to stay updated about this increasingly important marketing area, enter "optimizing search engine results for foodservice operations" in your favorite browser and view the results.

Social Media Sites

The rise in popularity of **social media** sites makes them an essential tool for marketing foodservice operations. Just as operators can choose the content placed on their proprietary websites, they can also select the content used in their online social media accounts.

The use of social media sites for social networking continues to grow. Increasingly, they are the sources where consumers seek information and how they seek information about foodservice operations they may want to visit.

Key Term

Social media: A collective term for websites and applications that enable users to create and share information and content or to participate in social networking.

The Importance of Social Media

The extensive use of social media is relatively new. Facebook was started at Harvard in 2004, and by 2020 it boasted 2.8 billion monthly active users. Pinterest was started in 2010, and as of 2022, this site reported nearly 500 million users worldwide.

The increase in popularity of social media sites is due to several important characteristics:

1) The sites are user-friendly: It takes very little technological knowledge to post content on a social media site. That means that most of an operator's customers can easily participate in social media regardless of the demographics of their target markets.
2) The sites provide around-the-world networking opportunities with real people, and the introduction of profiles on social media sites allows people to learn information about other persons before they interact with them. The primary purpose of social media accounts is communication and interaction, and a social media account differs from a website. A traditional website provides communication that is one-way, and social media users primarily seek two-way communication.
3) The sites allow the creation of groups: The most popular social media sites allow users to create groups with like-minded people who share similar interests, hobbies, or experiences. These groups act like clubs whose membership is open to anyone who shares these common traits. And, like most clubs, those who participate most often receive the most value from the membership.
4) The sites are free to use: Social media sites do not charge their users; rather they generate their income through the sale of ads, and this maximizes their accessibility.
5) The sites post reviews of businesses. In nearly all cases, customer reviews of a business posted online are given more weight than advertisements placed by

the same business. User-generated reviews of a business can increase or decrease the credibility of the business's products or services in the minds of potential customers before they make a purchase decision.

Social media sites are important marketing channels (see Chapter 4) for food-service operators. One important reason is that they allow the operation to reach large and targeted audiences both through their own social media account postings and with paid advertising.

Posting Content on Social Media Sites

It is estimated that one out of every four people uses social media daily and that 4.2 billion of those users access information from hand-held computer devices. Foodservice operators who wish to establish an online social media presence to reach these users should select one or more social media sites and spend their time focusing on quality posts, conversations, and ways in which to foster engagement with potential customers.

Foodservice operators must create their own social media accounts before they can post desired content on a social media site. Doing so is usually easy and, in most cases, entails no costs other than the time involved to create and maintain the account. The popularity of any specific social media site may increase or decrease over time but, at the time of this writing, the important sites for operators to consider joining are:

✓ Facebook
✓ YouTube
✓ Twitter
✓ Instagram
✓ Pinterest
✓ TikTok

Facebook

Its sheer size (Facebook accounts for nearly 25% of all Internet traffic) is one reason operators should establish a Facebook account. Doing so allows operators to create their own Facebook "page" to describe their business. It also enables operators to post pictures, videos, and menu updates which can generate strong relationships with other Facebook users.

YouTube

YouTube allows operators to post videos of their business, and most Internet users prefer to learn information from videos rather than written communication. YouTube videos allow operators to share images about new menu items, showcase a new chef joining the operation's team, or demonstrate how menu items are prepared. In most cases, an acceptable video for posting can be made using only a smartphone and its built-in video editing features.

Twitter

According to the IACP Center for Social Media, as of 2020, there were more than two billion searches on Twitter each day. This site allows operators to post updates consisting of 140 characters or less to promote their events, menu items, and specials. On this site, operators can also create memorable "hashtags" to use on their posts to help customers find them in the future (examples: #Tonyspizza" or #Trytonyspizza").

Instagram

Instagram was created to allow users to post interesting photos. This mobile photo app is especially popular among younger users. Foodservice operators can post their own photos and motivate their guests and followers to do so with photo contests. For example, a guest could win a prize package by taking a picture of themselves enjoying menu items the operator most wants to promote. An operation's Instagram "profile" can be linked to their proprietary website and other social media accounts.

Pinterest

Much like using a corkboard, Pinterest lets users "pin" their favorite findings on the Internet with links and images. Food is the top-ranked search category on Pinterest. One reason: Many users pin recipes on Pinterest and view Pinterest photos of food items other members have prepared or ordered and consumed at their favorite restaurants.

Some operators create Pinterest boards that break down the categories on their menu and make, for example, separate boards for appetizers, salads, entrees, and desserts. Then they take pictures and post photos of each of these items so future customers can see exactly what they will receive when they place their orders.

TikTok

TikTok provides a good example of why foodservice operators must continually monitor the changing popularity of social media sites. It was started in late 2018, and within three years had more hits per day than Facebook. TikTok can be thought of as an "appetizer" version of YouTube because the TikTok app allows viewers to post videos 15 seconds in length (or shorter). It is especially popular with Millennials (those born between 1981 and 1996) and Generation Z (those born between 1997 and 2012).

Operators utilizing TikTok can connect short video clips together, incorporate music, and add filters, green screens, effects, stickers, and more to make their clips engaging. For example, a pizza shop operator could share videos of their pizzas as they come fresh out of the oven and are being portioned. The video could then instruct viewers to find more information about the business on the operator's proprietary website.

Find Out More

Foodservice operators seeking a current book about "Social media marketing for foodservice operations" will likely be disappointed. A reason: The popularity of social media sites and the ways they can be used change extremely rapidly. As a result, operators are more likely to find up-to-date and useful information by searching the daily postings of the hospitality trade press for specific articles about the use of an individual social media site.

It is important to recognize that not every social media site is suitable for every type of operation. For example, a private Country Club that primarily caters to its older members would not find their marketing efforts using TikTok to be as effective as other social media sites that are more popular with older generations of users. Therefore, operators should learn the most important marketing-related information about the social media sites that are of the most value to them.

To stay up-to-date on the changing ways that social media platforms can be used to market foodservice operations, select a popular media site, and then enter "(*site name*) marketing for restaurants" in your favorite browser and view the results.

This chapter has addressed how foodservice operators can increase their business and their online presence using proprietary websites and the content they post on social media sites they join.

There are, however, other sites on which information about a foodservice operation can be posted by other businesses and business partners. In most cases, an operator cannot readily change this information. Online content about an operation posted on third-party managed sites is the sole topic of the next chapter.

What Would You Do? 9.2

"I think you have your priorities all screwed up," said Otis.

"What do you mean by that?" asked Samantha.

Otis was Samantha's grandfather. He had been the owner and operator of Otis Candies, a small retail candy store that specialized in hand-crafted artesian chocolates. The shop had been operating for over 40 years, and Samantha was spending her summer off from college helping Otis in the store. Thirty days after her arrival, online orders processed on the store's website had increased by nearly 50%.

"What I mean is we should be spending our time speeding up the processing of all of our online orders. That's where our business is increasing the most. I think you're wasting your time with all of these social media accounts that you started," said Otis.

(Continued)

Otis was right that Samantha had established an online presence for Otis Candies with three of the most popular social media sites. She enjoyed posting information and short videos about the shop's long history and how it makes its popular gourmet chocolate items.

"Well, maybe using the social media accounts to drive business to our website for ordering is why it's been so busy lately," replied Samantha with a smile.

Assume you were Samantha. How could you go about providing Otis with information about the impact of a social media presence on the increase in online orders? How important would it be for her to do so?

Key Terms

Domain name

IP address

Extension (domain name)

Registration
   (domain name)

Web host

PDF

Proprietary website

Google search

F-shaped scanning
   pattern

Navigation (website)

User experience

Traffic tracking (website)

Click-through rate (CTR)

Bounce rate

Search engine results
   page (SERP)

Search engine
   optimization (SEO)

Keywords

Social media

Operator's 10-Point Tactics for Success Checklist

Evaluate your need for, and the current status of, each of the following operational tactics. For those tactics you think are important, but not yet in place, develop an action plan for its implementation including who will be responsible for the tactic's completion and the target date by which it should be completed.

Tactic	Don't Agree (Not Done)	Agree (Done)	Agree (Not Done)	Who Is Responsible?	Target Completion Date
				If Not Done	
1) Operator understands the importance of choosing a primary domain name that will drive traffic to the operator's proprietary website.	___	___	___		

Tactic	Don't Agree (Not Done)	Agree (Done)	Agree (Not Done)	If Not Done	
				Who Is Responsible?	Target Completion Date
2) Operator has carefully considered the important factors to be evaluated when choosing a website host.	___	___	___		
3) Operator understands the importance of having a professionally developed proprietary website.	___	___	___		
4) Operator understands the importance of professional website design when creating a proprietary website.	___	___	___		
5) Operator understands the importance of website content when creating a proprietary website.	___	___	___		
6) Operator understands the importance of tracking traffic on their operation's proprietary website.	___	___	___		
7) Operator has a basic understanding of how a search engine results page (SERP) is generated.	___	___	___		
8) Operator understands the importance of search engine optimization (SEO) efforts related to their proprietary website.	___	___	___		
9) Operator recognizes the increasing importance of social media in the overall marketing effort.	___	___	___		
10) Operator understands the value of regularly posting and updating content on their chosen social media sites.	___	___	___		

10

Web-Based Marketing on Third-Party Sites and Apps

What You Will Learn

1) The Importance of Local Links
2) How to Utilize Third-Party Website Partners
3) The Advantages of Third-Party Delivery App Partnerships
4) The Disadvantages of Third-Party Delivery App Partnerships

Operator's Brief

In the previous chapter, you learned that a foodservice operator's proprietary website is usually its single most important marketing tool. In this chapter, you will learn about ways content from an operator's proprietary website can be viewed on the websites of other entities.

There are several reasons why linking with other websites can be very helpful. First, doing so can increase the visibility of an operator's proprietary website when it is linked with other frequently viewed websites. When this is done, visitors to another website can click on the operator's link and be directed to their operation's website.

When an operation creates a significant number of links with others who operate websites in a local community, the exposure gained can be very significant. This is especially so if the local links are thoughtfully created with website operators whose products and services are complementary to those of the foodservice operation.

A second way to increase the visibility of an operation's website is by partnering with local and national third-party website operators. Typically, there will be some restrictions placed on the format or content third-party website operators permit on their sites. However, visitors to the third-party sites will be able to view important information about a foodservice operation and, sometimes, gain the ability to link directly to the operator's proprietary website.

Increasingly, some foodservice operators consider creating partnerships with online companies offering third-party delivery services. There are advantages and disadvantages to these partnerships, and both are addressed in this chapter. The chapter concludes by addressing issues of self-operated delivery. While this approach also has advantages and disadvantages, it is increasingly a topic foodservice operators might consider as they market professional meal delivery services to their guests.

CHAPTER OUTLINE

The Importance of Local Linkage
Marketing on Third-Party-Operated Websites
Partnering with Third-Party Delivery Apps
    Advantages of Third-Party Delivery App Partnerships
    Disadvantages of Third-Party Delivery App Partnerships
Self-Delivery Options

The Importance of Local Linkage

One good way for foodservice operators to expand the visibility of their proprietary websites is by linking their own sites with other website operators in their local communities. Linking is accomplished by providing a website viewer with a **link** to the operation's own website. A link, short for "hyperlink," is simply a connection from one webpage to another.

Links are used when a website owner wants to connect their webpage to that of a partner organization, client, friend, blogger, or hobby group. For foodservice operators with a professionally developed and effective proprietary website (see Chapter 9), links can be an excellent way to expand their online presence and reputation.

Links can be found on almost every webpage, and they provide a simple means of navigating between pages on the web. While links can connect one website to another, they can also be used to go from one page of a site to another page on the same site.

The type of link typically of great importance to a foodservice operator is a **local link** that connects the web content of a foodservice operation

Key Term

Link: A means of easily navigating from one webpage to another or from one section of a webpage to another section of the same webpage.

Key Term

Local link: Website links created to show that other entities with relevance in the local area trust or endorse a foodservice operation.

with that of another locally operated webpage. The purpose is to show that another organization with significant relevance in their local area trusts or endorses the operator's business. When the visitors to a local website click on the link, they will immediately see the content presented on a foodservice operator's proprietary website because they will be transported to that site.

To illustrate how a local link can benefit both linking partners, assume that a foodservice operation is located within one mile of the city zoo. It is possible that some visitors to the zoo might look for a foodservice operation close to the zoo for food and/or beverages either before or after the visit to the zoo.

Assume, when reviewing the zoo's website to learn about operating hours and driving directions, potential visitors to the zoo also saw a link to the foodservice operation's webpage. Some of these potential visitors might also click on the operation's link to gain foodservice information, and, as a result, the operation gains potential customers.

As a second example, assume that a foodservice operation is located very near a limited-service hotel that does not provide on-site food options for guests. It is likely that the hotel would want to create a local link with the foodservice operation to inform hotel guests about convenient foodservice options in the local area. In return, the foodservice operation's link to the hotel's website could also help in giving directions to it. (Example: "We are located right across the street from the Comfort Inn on West Saginaw Highway.")

Operators should recall that creating local links to popular websites also help a foodservice operation improve its ranking on search engine results pages (SERPs); see Chapter 9. The primary purpose of a link is to increase the visibility of the operation's proprietary website, and there are several ways foodservice operators can do this:

1) Linking with other businesses

  Effective link building is about building relationships. Those businesses who operate their own websites are more likely to add a link to a foodservice operator's website if they have a genuine connection to the operator.

  Foodservice operators should research local businesses that provide products and services that complement those offered by the operator. For example, a foodservice operator may partner with a local taxi service by adding a link to the taxi company's site, and the taxi company could provide a link back to the operator's website. Both businesses would benefit from the referrals resulting from this linkage.

2) Sponsoring local events

  In many cases, local community events depend on sponsorships to cover the event's costs. In exchange for the sponsorships received, the event's organizers may provide promotional benefits including website links to the sponsor's business.

Foodservice operators can monitor their local areas for conferences, concerts, festivals, and sporting and professional events held within their target market area. Which are likely to attract people who are potential customers? Any event with significant numbers of attendees who could be part of a foodservice operation's target market may be a good fit and could be given a sponsorship.

3) Sponsoring local organizations

In most communities, there are large numbers of nonprofit organizations and charities that seek support from community business leaders. Sometimes, linking with these organizations is a better tactic than sponsoring a local event because links with nonprofit organizations are continuous instead of only a few days or weeks.

For many foodservice operations, logical links with local nonprofits include organizations whose goals address healthy eating and living, green practices (see Chapter 2), and providing food and shelter to those in need. Local school systems may also seek sponsorships for items and events for which external funding is needed. Examples include sponsorships addressing needed teaching supplies, sporting events, and sponsoring music and arts programs.

4) Joining appropriate professional organizations

Nearly all cities offer operators several professional organizations including membership in a local area's Chamber of Commerce, Restaurant Association, or **Convention and Visitors Bureau (CVB)**. These opportunities allow operators to interact with other local professionals and, in many cases, a business can link to a membership directory page found on the organization's proprietary website.

Key Term

Convention and Visitors Bureau (CVB): The local entity responsible for promoting travel and tourism to and within a specifically designated geographic area.

In some cases, professional organizations may be based on demographic factors. An operator may elect to support a Hispanic Chamber of Commerce, a Small Business Chamber, an LGBTQ Chamber, a Professional Women's Chamber of Commerce, and others.

5) Connecting with local bloggers

Most mid-size to larger communities have local bloggers who write about things to do in the area or about issues relevant to those living in the community. These blogs may provide foodservice operators with an opportunity to create mutually beneficial links. For example, an operator might invite the blogger to visit the operation and sample menu items at no charge. In exchange, the blogger might write and post a positive review and place a link to the operator's proprietary website on their blog. Additionally, an operator might offer a special promotion for anyone who places an order and states that they heard about the special promotion after first visiting the blogger's site.

6. Participating in local awards programs

In many communities, local media including radio, television, newspapers, and bloggers regularly conduct "Best of the Area" awards programs. These programs typically have a large number of categories but almost always include foodservice operations in multiple categories, for example, "Best Italian," "Best Pizza," best "Food Truck," and the like.

High rankings in awards programs can result in an operation's website being promoted or listed and then linked with the website announcing the awards.

Find Out More

Link building is an important part of search engine optimization (SEO) efforts. The larger the number of high-quality sites that mention a foodservice operation's name and provide a link to its website, the higher the operation will rank in SERPs.

While it does require some effort, in nearly all cases foodservice operators will find they have opportunities to link their own sites with many others in their local communities.

To learn about some ways foodservice operators can improve their web presence by linking their websites with local organizations and businesses, enter "link building strategies for restaurants" in your favorite browser and view the results.

Marketing on Third-Party-Operated Websites

Foodservice operators can deliver information from (and in some cases create links to) its proprietary website to other websites by placing some of its webpage content on one or more third-party operated sites. Note: Unlike an operator's website, third-party-operated websites control the content they include on their sites. In some cases, the content on third-party websites is supplied directly by an operator and, in other situations, this will not be the case.

One example of a third-party content-controlled website providing information about a foodservice operation is a "Google" search. In Chapter 9, you learned that a SERP lists results returned by a search engine in response to a user's specific word or phrase query. Figure 10.1 shows a portion of the SERP produced when the foodservice operation "Margaritas To-Go, Houston" is entered into Google search.

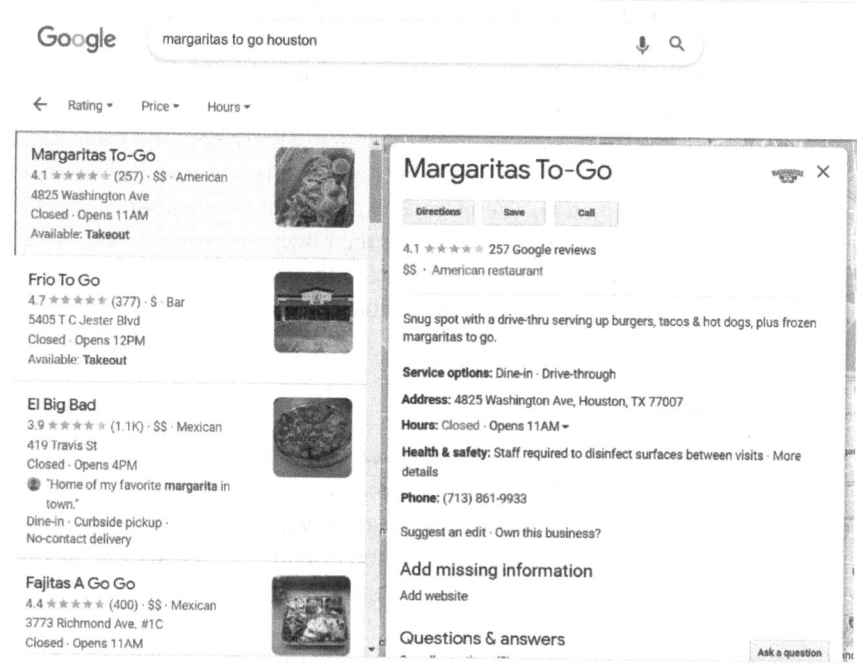

Figure 10.1 Google Search Results

The Google search pages show Margaritas To-Go listed first on the SERP. In this example, Google has provided a significant amount of information about the operation including its name, address, and phone number (NAP). Customer reviews have been posted along with hours of operation, and space is provided for the operation to add its website as a link, if desired.

Note also that Google search has listed (left side of the results page), other food-service operations that might be of interest to the viewer/searcher.

The SERP Google generated for this search included, among many others, the additional third-party-operated websites listed in Figure 10.2.

The list above is not exhaustive, but it does illustrate some of the different types of third-party-operated websites listed on the SERP for this search. It is important to understand that all these sites (and others!) provide information about this foodservice operation. In some cases, the information provided on these third-party-operated sites may be accurate and up-to-date. In other

https://www.facebook.com
https://365thingsinhouston.com
https://houston.eater.com
https://papercitymag.com
https://www.visithoustontexas.com

Figure 10.2 Selected Websites Displayed on "Margaritas To-Go" SERP

instances, it may not be. In all cases, the information will be displayed in a format determined by the third-party website operator.

Regardless of the information listed on the third-party site, a foodservice operator would not typically be allowed to go into these sites to correct erroneous information. Instead, the third-party website operator wholly controls each site's content, and he or she may or may not provide an opportunity to easily correct or modify erroneous information. Therefore, it is important that foodservice operators routinely undertake Internet searches for their own businesses to learn what the SERP shows, and to then follow up with the third-party website operator to correct dated or otherwise incorrect information.

What Would You Do? 10.1

"I'm looking at your menu right now, and I want a take-out order of the barbeque beef ribs!" said the frustrated guest on the phone.

The guest was talking to Jeremiah Larson, the owner of Jeremiah's Backyard Barbeque and Brew, an operation that featured barbecued meats and sides such as potato salad, baked beans, macaroni and cheese, and an extensive selection of craft beers.

"I'm sorry, Sir," said Jeremiah, "We took the barbecue beef ribs off the menu last year. We had trouble ensuring a good supply of high-quality beef ribs, so we replaced that item with Brisket Burnt Ends, and I think you would enjoy them very much."

"Well, if you took them off the menu why are they still on the menu shown on the "Taste of Our Town" website? I'm looking at a photo of your menu, or maybe it's a PDF copy of your menu, right now!"

Jeremiah was familiar with the website and knew its operator routinely received cell phone photos of menus taken by diners at various local operations. The diners could then upload photos to the website's review section where they were posted for viewing by all visitors to this popular local dining guide's site. This user-supplied content made it easy for the operator of the "Taste of Our Town" website to show what local restaurants were serving. Likewise, they could do so without the cost and efforts required to re-format and post the printed menus used by many different properties.

However, the site was now causing a big problem because it showed a picture of last year's—not this year's—menus!

Assume you were Jeremiah. How important would it be to regularly monitor third-party websites listing information about your restaurant to help ensure the information was accurate? What would be a good way to implement a regular monitoring program?

Technology at Work

Foodservice operators wanting to gain greater visibility in Google searches can create a Google business listing officially known as a "Business Profile." However, developing this tool does not allow an operator to manage and edit information in the profile. To do that, operators must, in addition to creating their business profiles, establish a "Google Business Profile" account.

There is no charge for the accounts, which are used to manage and edit the content contained in an operation's business profile. The account also allows an operator to interact with those who view the operation's business profile.

With a Google Business Profile account, operators can respond to reviews, answer questions, and send direct messages. They can also utilize their account to publish posts to their business profile in much the same way as they post content to their Facebook and other social media accounts.

To learn more about how foodservice operators can utilize this free technology tool to enhance their Google search results, enter "setting up a Google Business Profile account" in your favorite browser and view the results.

The results shown from a Google search provide a good example of content displayed on third-party-operated sites. The content placed on an operation's Facebook page is another good example. Recall from Chapter 9 that foodservice operators can establish their own accounts on numerous social media sites.

Facebook is a very popular site, and many foodservice operators establish their primary online presence with this social media website. Note: in some cases, operators may even create a vibrant Facebook page instead of a proprietary website. One reason: the SERP for their operation listed "Facebook" (https://www.facebook.com) as one of the top sites. Note: the reason for the popularity of the Margaritas To-Go search (see Figure 10.2) was because that property created a significant online presence with the Facebook page it maintains. That presence, in turn, influenced the SERP generated from the Google search.

Facebook, a third-party-operated site, provides a good illustration of how operators can increase their online presence through partnerships with well-managed third-party-operated sites. Some content suggestions that could be placed on Facebook's third-party-operated site (as well as other sites that permit it) include:

1) Posting menu item information and photos

For most foodservice operations, posting information on third-party-operated websites is a good way to increase the visibility of the menu items they sell. Operators can easily take high-resolution photos of the dishes they sell and publish them once or twice a week on their Facebook page. This SEO strategy works because SERPs are directly influenced by the frequency with which new information is placed on a website.

Find Out More

A professional food photograph is a powerful marketing tool. It can make viewers hungry, inspire them to make a reservation, or even convince them to immediately place an online order for the item.

Professional food photographers know that there are some simple steps foodservice operators can use to help ensure that food photographs look as enticing as possible. Some of these professionals have placed tips on YouTube that can be very instructive.

To see some suggestions for enhancing the quality of food photographs, go to the YouTube website, enter "how to photograph food" in the search bar and view the results.

---

2) Showing the operation

Foodservice operations that offer on-site dining and that show pictures of their dining areas allow viewers on third-party-operated websites to visit the restaurant virtually. Photos of the operation's exterior entrance doors, dining and lounge areas, and table settings can all increase viewers' comfort levels because they can see online what they will experience when they visit the operation.

3) Showing the operation's staff

With the permission of staff members, operators can post photos of their managers, servers, and kitchen staff on their Facebook pages. Pictures of smiling staff members wearing their uniforms can help put a personal "Face" on the operation.

4) Showing the operation's staff in action

Posting pictures of staff members serving happy guests or performing interesting kitchen tasks is typically a good idea for a Facebook posting. If operators utilize this marketing tactic, they must ensure that the staff members are demonstrating safe food practices and that any background areas shown in the photos are clean and professionally maintained.

5) Provide interesting food-related information

Guests viewing content posted on third-party-operated sites are often interested in learning more about food and its preparation. Facebook postings can describe unique menu ingredients and where they are from, or unique preparation methods and why they are used. Both tactics help educate customers and enhance the personal relationship between the site's viewer and the operation.

6) Sharing positive reviews

Facebook pages allow guests to post reviews of the foodservice operations they have visited. When the reviews are positive, they can be highlighted and shared on the operation's initial Facebook page. These postings might include excerpts of positive reviews and a management reply thanking reviewers for sharing their opinion. Facebook page reviewers will also better understand that they will also receive a positive experience when visiting the operation.

Foodservice operators desiring to establish partnerships with third-party-operated websites will find they have many opportunities to do so. Each third-party website operator may have their own rules and procedures for posting content. Foodservice professionals should select the third-party-operated sites they feel will be most useful to their operations. They can then make maintaining a relationship with those sites an important part of their annual marketing plans.

Partnering with Third-Party Delivery Apps

Foodservice operators make important marketing-related decisions as they consider whether to partner with one or more businesses operating third-party delivery apps (see Chapter 1). Companies that have created third-party delivery apps can make it easy for foodservice guests to place online orders for meal delivery to their homes, offices, or other desired locations.

Many foodservice professionals agree that the first third-party delivery app was offered by World Wide Waiter (now operating as "Waiter.com") in 1995. During the early 2000s, food delivery dramatically increased in popularity. As smartphone usage became commonplace, office workers found they could order food from their phones to the office without having to step away from their desks. In addition, consumers who preferred not to cook for themselves after work found it was easy to order home-delivered meals with a few taps on their smart devices.

The COVID pandemic of 2019 again significantly increased the use of third-party delivery apps: many foodservice operations were closed for indoor dining and, also, many guests decided it was safer to eat in their homes than dine out. Today, DoorDash, Uber Eats, Postmates, and Grubhub are among the largest and most popular third-party delivery companies having developed apps. Each of these companies exhibits strengths in some geographic markets and has a lesser impact in other areas.

The basic business model of third-party delivery companies is relatively straightforward. Foodservice operators agree to have their operations listed on the app, and the operator then provides a menu that includes item prices. Guests utilizing the apps place their orders with the third-party delivery company (not directly with the operation).

Orders placed on the app are electronically transmitted to the foodservice operation's point-of-sale (POS) system. The third-party delivery company uses a cadre of independent drivers to pick up the orders and deliver them to guests. In addition to paying for the food they have ordered, customers also pay a delivery fee to the third-party delivery company. Drivers are paid on a per order basis by the company, and they typically retain any tips generated by the orders they have picked up and delivered.

The third-party delivery company charges a commission to the foodservice operation for their delivery services. Typical commission fees range from 20% to 30% of the dollar amount of the order. The remaining amount of the food order's value is then transferred to the foodservice operation's bank account. Foodservice operators, whether traditional or ghost (see Chapter 1), may partner with only one or with multiple third-party delivery services.

Advantages of Third-Party Delivery App Partnerships

Unless required as part of their franchise agreement (as negotiated by the franchisor), foodservice operators have the discretion to create partnerships with third-party delivery organizations.

Third-party delivery app operators aggressively recruit restaurants to join their sites. In fact, during the earlier days of third-party delivery app expansion, some third-party app operators listed restaurants on their sites who had not formally agreed to a partnership. This practice is no longer utilized by third-party apps, but it does speak to the aggressive nature of some third-party delivery companies in expanding their networks.

The advantages of partnership expansion to a third-party delivery operator are easy to understand. The more foodservice operators listed on the third-party site, the better the chances that customers will find an operation from which they wish to order, and the higher will be the commissions earned by that organization.

As they market apps to foodservice operators, third-party delivery companies generally stress three key factors to encourage foodservice operation sign-ups:

✓ App popularity
✓ Competitive pressures
✓ Impact on profits

App Popularity

Most foodservice professionals agree that how the "dining out" world is eating is dramatically changing. As recently as 20 years ago, restaurant meal delivery services were largely limited to selected industry categories such as pizza and Chinese food. Also, operators offering delivery services typically employed their own staff to transport food to the consumers.

Today, food delivery is a global market generating more than 150 billion in sales annually, having more than tripled since 2017. Meal delivery in the United States doubled during the COVID-19 pandemic, and industry growth has slowed somewhat from its pandemic heights. However, expansion continues to be significantly strong, and 80% of Americans have ordered food for home delivery. It is now estimated that, by 2023, there will be more than 55 million users of third-party delivery apps and, by 2025, food delivery will comprise 21% of all commercial foodservice operation sales.[1] Third-party delivery app operators point to the increasing popularity of home meal delivery as a reason foodservice operators should join them. This reasoning is certainly supported by data related to changes in the market behavior of today's consumers.

Competitive Pressures

A second major reason frequently cited by third-party app operators seeking partnerships is that a foodservice operation's direct competitors will be listed on their sites. As a result, some foodservice operators reason that, since their competitors will be listed on the site, they must also be listed to maintain competitiveness.

Using this argument, third-party app operators point to an increasing number of customers who desire online ordering and delivery option(s). If this service is not offered, some potential customers will select a different restaurant that does offer delivery, and the operator will lose the chance to get these customers' sales.

This argument gains some credibility over time. Consider 2020, when DoorDash formed a partnership with Little Caesars pizza, a brand that had never offered off-site delivery. Note: DoorDash also has partnerships with Wendy's, Chick-fil-A, and McDonald's, the biggest fast-food chain in the country. McDonald's also partners with Uber Eats. At the time of this book's writing, Starbucks had a contract with Uber Eats, Popeyes used Postmates for third-party delivery, and Taco Bell and KFC maintained a third-party agreement with Grubhub.

Impact on Profits

Those foodservice operators who partner with third-party delivery services do so with the belief their profitability will increase. In fact, these third-party delivery companies often indicate that cost savings do accrue to foodservice operations that do not need to create a unique online order and delivery system because it is provided by the third-party service.

As previously indicated, foodservice operators can incorporate online ordering and delivery service into their own websites (see Chapter 9). However, there is a cost to do so. In addition to setting up an online ordering system, operators implementing their own delivery service incur additional costs of operating its delivery fleet and paying its drivers. Therefore, partnering with a third-party delivery organization will likely save operators these costs as well.

---

1 https://beambox.com/food-delivery-service-statistics. Information retrieved on August 15, 2022.

Third-party delivery app operators also suggest that the typical online delivery order is larger than that typically placed by on-site diners, so the guest check average is higher with online delivery services. Also, since third-party delivery operators charge commissions on each sale, the foodservice operation only pays a fee when a sale is generated.

Disadvantages of Third-Party Delivery App Partnerships

Despite their increasing popularity, some foodservice operators point to perceived disadvantages of developing partnerships with third-party delivery companies. While the precise reasons vary, operators who do not strongly support third-party delivery app partnerships typically point to three areas of concern:

✓ Quality control
✓ Guest ownership
✓ Financial concerns

Quality Control

When partnering with third-party app delivery companies, foodservice operators have 100% control over ordered menu items as they are produced and packaged for delivery. Once the food is picked up for delivery, however, operators have virtually no control if challenges arise.

For example, if it takes a long time for a delivery driver to deliver an order, there is little the foodservice operator can do. Similarly, if a driver's appearance is unprofessional, he or she "samples" the guest's order, smokes, or makes deliveries with pets in his or her vehicles, the third-party service must address these or related issues. However, third-party apps typically do not employ their own drivers. Instead, they partner with independent contractors who make the food deliveries, and it can be challenging for third-party app operators to enforce driver performance standards.

The issue of driver control becomes even more controversial when guests accept the food and rate the delivered meal to be of poor quality because of the driver's performance. What should occur if a delivery driver refuses to enter an elevator for a 10th floor delivery to a guest and, instead, requires the guest to be in the lobby to pick up the order? The guest may give a poor rating to the restaurant even though it was the driver's performance, not that of the restaurant, which was inadequate. Given the critical importance of guest ratings posted online, poor reviews resulting from improper delivery service can be very damaging to a foodservice operator's reputation.

As a final aspect of this area of concern, foodservice operators point out guests' primary complaint with delivery services is the slow length of service. This can happen during extremely busy times, but it can also occur because of how delivery drivers are compensated. To illustrate, assume a guest has placed an order

requiring a 15-to-20-minute trip from the time of pick up at the foodservice operation to the time of delivery to the guest (a 30-to-40-minute round-trip). Assume
further that the guest has placed an order with a dollar value of $100 but has
chosen to tip only $2.00. In this example, it is highly unlikely that a driver would
be motivated to deliver such an order in the future. As a result, the phrase "No Tip,
No Trip" is now well known among delivery drivers and foodservice operators.

Each third-party delivery app has incentive programs to motivate their drivers
to take an order, even without a tip. In many cases, the delivery app operator
increases a driver's payment fee either if an order is small (and the tip is also likely
to be small) or in which the tip amount is unknown. This topic of equating quality
delivery services with tip size when partnering with third-party app operators is
one that foodservice operators must carefully monitor.

Find Out More

It is important for foodservice operators to recognize that those companies
offering third-party delivery of the operators' menu items do so with "employee
drivers" who are not actually employees. Rather, the third-party delivery drivers are independent contractors. An independent contractor is a person that
performs services for another person or entity under a contract between them,
with the terms of the relationship spelled out such as duties, pay, and the
amount and type of work to be done.

Because of the independent contractor relationship, a third-party delivery
company cannot *require* one of its contracted drivers to perform a service (i.e.
make a delivery). Each independently contracted driver can elect to accept, or
not accept, any order the third-party delivery company wants them to deliver.

Increasingly, those who drive for third-party delivery companies are selective in the orders they wish to deliver. The reasons why they are selective are
economic. For example, if a requested lunch delivery from a QSR has a price
tag of $6.00, a delivery driver will recognize that the gratuity likely to be paid
on such an order will be very small. As a result, a driver may simply elect to
refuse to deliver that lunch order, while waiting for a delivery request for a
larger order.

Similarly, if a delivery order is small, but the distance required to deliver the
order is great, an independent contractor can quickly calculate the cost of fuel,
vehicle wear and tear, and the time required to make the delivery, and elect to
decline delivery of the order.

Increasingly, there are widely reported incidences of multiple smaller
orders remaining in a foodservice operation because no driver is willing to
deliver them. Foodservice operators should continuously monitor this situation as it is likely to continue to increase in importance in the future. To do
that, enter "current delivery driver employment issues" in your favorite browser
and view the results.

Guest Ownership

Guest ownership is a significant issue for operators that are wary of third-party delivery apps. If a guest uses an in-house operated online ordering system or personally calls in an order, the customer could be enrolled in and earns points from the operation's customer loyalty program (see Chapter 8). This, in turn, encourages repeat business, and information about the customer such as credit card data, e-mail address, and a telephone number could be retained by the operation and used for other marketing purposes.

Guests utilizing a third-party delivery app, however, are technically customers of the third-party delivery app company. As a result, vital guest information including size of order, frequency of ordering, and preferred payment method will be retained by the third-party app, and this important data will not likely be shared with the foodservice operator.

Third-party delivery apps are interested in keeping their own customers happy. Problems can occur if, for example, a customer calls a third-party delivery company to complain about a missing side dish or lack of condiment that was ordered. If the delivery company decides to refund the guest's entire bill, the foodservice operator will have incurred the cost of food and labor to produce the order but received no revenue from the sale.

Financial Concerns

One complaint that foodservice operators (and some guests!) have with third-party delivery app operators involves delivery and service fees that must be paid. For foodservice operators, the delivery fees paid to third-party delivery app operators reduce the amount of money that will remain to pay for food, labor, and other operating expenses required to produce the menu items delivered.

In many cases, after delivery fees are considered, foodservice operators may make very small profit margins, may break-even, or may even lose money on a delivery sale. Note: in nearly all cases, foodservice operators cannot charge higher rates for foods delivered by third-party operators than are charged to regular online guests.

To illustrate the financial impact of partnering with a third-party delivery app consider the data presented in Figure 10.3. Figure 10.3 is an **income and expense statement** prepared using the **Uniform System of Accounts for Restaurants (USAR)**.

Key Term

Income and expense statement: A detailed listing of revenue and expenses for a given accounting period. Also commonly referred to as an "Income Statement," "Profit and Loss Statement," or "P&L."

Key Term

Uniform System of Accounts for Restaurants (USAR): A recommended and standardized (uniform) set of accounting procedures for categorizing and reporting restaurant revenue and expenses.

Figure 10.3 consists of three columns of data. The first column indicates P&L results with $1,000,000 in annual sales with no third-party delivery partner relationship. The second column presents data if a third-party delivery partner was selected, and that partnership generated an additional 10% of the operation's current business. The third column represents the results that would occur if the third-party app produced an additional 50% of the operation's total business.

Foodservice operators making their own careful comparisons related to the financial impact of partnering with third-party delivery apps must make their own reasonable assumptions as they assess the impact of the partnership. The assumptions used in the income statement comparison shown in Figure 10.3 were as follows:

1) The cost of food percentage for on-site and delivered meals remain unchanged at 30% of sales.
2) Total delivery fees and commissions on delivered meals equal 27.5% of sales.
3) Total takeout packaging costs for delivered meals equals 5% of delivery sales.
4) Increases in management cost to supervise the increased sales levels are small but have been accounted for.
5) Variable labor (staff) costs remain unchanged at 30% of total sales.
6) Employee benefits remain unchanged at 17% of total labor costs (cost of Management + Staff).
7) Other controllable costs remain unchanged regardless of volume with the exception of changes in utility costs that are small but have been accounted for.
8) Non-controllable costs remain unchanged.
9) Interest expense remains unchanged.
10) Income tax rate of 20% remains unchanged.

Given the 10 assumptions cited before Figure 10.3, a careful review indicates that an increase in delivery sales of $100,000 (Column 2) or even $500,000 (Column 3) will have little impact on the operation's **net income**. Using the assumptions indicated, it is not too surpris-

Key Term

Net income: The profits achieved in a foodservice operation after income taxes have been paid.

ing given that, with a 27.5% delivery fee and a 5% packaging fee, the additional sales generated in this specific example make a negligible contribution to net income.

It is important to note that, to make a positive impact on its income, the foodservice operator in this example must either negotiate a reduction in commission and/or delivery fees, reduce their packaging costs, and/or find savings in other areas (such as reducing its marketing expense).

Casa de Queso

	Column 1	Column 2	Column 3
	No Third Party	Plus 10% Third Party	Plus 50% Third Party
SALES			
Food Sales	$1,000,000	$1,000,000	$1,000,000
Third-Party Delivery Sales		100,000	500,000
Total Sales	1,000,000	1,100,000	1,500,000
COST OF SALES			
Food @ 30%	300,000	330,000	450,000
Delivery fees @ 27.5%		27,500	137,500
Takeout Packaging @ 5%		5,000	25,000
Total Cost of Sales	300,000	362,500	612,500
LABOR			
Management	80,000	82,000	90,000
Staff @ 30%	300,000	330,000	450,000
Employee Benefits @ 17%	64,600	70,040	91,800
Total Labor	444,600	482,040	631,800
PRIME COST	744,600	844,540	1,244,300
OTHER CONTROLLABLE EXPENSES			
Direct Operating Expenses	40,000	40,000	40,000
Marketing	50,000	50,000	50,000
Utilities	10,000	10,100	10,500
Administrative & General Expenses	20,000	20,000	20,000
Repairs & Maintenance	5,000	5,000	5,000
Total Other Controllable Expenses	125,000	125,100	125,500
CONTROLLABLE INCOME	130,400	130,360	130,200
NON-CONTROLLABLE EXPENSES			
Occupancy Costs	10,000	10,000	10,000
Equipment Leases	0	0	0
Depreciation & Amortization	5,000	5,000	5,000
Total Non-Controllable Expenses	15,000	15,000	15,000
RESTAURANT OPERATING INCOME	115,400	115,360	115,200
Interest Expense	12,000	12,000	12,000
INCOME BEFORE INCOME TAXES	103,400	103,360	103,200
Income Taxes @ 20%	20,680	20,672	20,640
NET INCOME	$ 82,720	$ 82,688	$ 82,560

Data from Income statement prepared using current (8th. Edition) Uniform System of Accounts for Restaurants (USAR).

Figure 10.3 Sales Comparison Income Statement

In addition to fee-related concerns, some operators indicate that sales achieved with third-party delivery apps result in longer credit card processing times. Typically, foodservice operators can expect fund deposits from their **merchant services provider** every one to two business days. With third-party delivery app sales, the third-party entity will take out their commissions before depositing the remaining funds in the operation's bank account. This typically happens only once per week. If **cash flow** is an issue for an operation, an extended turnaround time for payments can present a significant disadvantage.

Self-Delivery Options

Given the various advantages and disadvantages of partnering with third-party delivery apps just addressed, foodservice operators generating a significant amount of delivery business should also consider the advantages and disadvantages of implementing a self-delivery program. While there are certainly costs related to doing so, there might be significant advantages to implementing a self-delivery program that include:

✓ Quality control of delivered foods
✓ Ability to set delivery charges
✓ Use of branded uniforms and transportation vehicles
✓ Increased control of guest data
✓ Ability to interface with guest loyalty programs
✓ Elimination of commission and fees
✓ The ability to utilize current staff for deliveries during slow periods
✓ The creation of jobs
✓ Increased profits

Key Term

Merchant services provider: A company that enables foodservice operators (merchants) to accept credit and debit card payments and other alternative payment methods. Also sometimes referred to as a "payment services provider."

Key Term

Cash flow: The net amount of cash and cash equivalents transferred in and out of a company. Cash received represents inflows, and funds spent represent outflows.

Find Out More

Domino's Pizza, headquartered in Ann Arbor, Michigan, is the world's largest pizza chain. As of 2023, the company franchised or operated more than 16,000 stores in 85 countries. Pizza is the company's primary focus, with traditional,

(Continued)

specialty, and custom pizzas available in a variety of crust styles and toppings. While store menus vary somewhat by location, in most stores additional entrees include chicken wings, pasta, bread bowls, and oven-baked sandwiches.

The typical Domino's Pizza location averages over $1,000,000 in sales per year. For franchisees, the royalty fee is 5.5% of the store's weekly sales, and the advertising payment is 4% of the store's weekly sales. These statistics help confirm that the outlets are popular with guests and profitable for operators.

Domino's promises fast delivery and at one point in its history (1973) actually guaranteed guest orders would be delivered in 30 minutes or less (a practice it has since discontinued due to safety concerns). The company is innovative in numerous ways, and, in 2021, it began testing self-driving delivery vehicles. In fact, Domino's does *not* now, nor has it ever, partnered with third-party delivery companies. Instead, the company hires delivery employees and supplies necessary delivery vehicles.

For operators considering launching or continuing their own delivery service, the Domino's perspective on third-party delivery partnerships may be enlightening. To learn more about the Domino's approach to home meal delivery, enter "Domino's home delivery innovations" in your favorite browser and view the results.

Technology at Work

One important challenge faced by foodservice operators considering self-delivery involves managing their delivery fleets. This is especially so for operators who do a significant amount of delivery business.

Fortunately, several companies have designed software for tracking foodservice deliveries. Operators can track driver locations, monitor the status of current orders, and provide estimated times of arrival (ETAs) to guests. The programs also track important delivery metrics such as average delivery time.

To learn more about the software programs foodservice operators can use to manage their delivery programs, enter "self-delivery assistance for restaurants" in your favorite browser and view the results.

The controversies related to the pros and cons of partnering with third-party delivery apps will no doubt continue. Also, some foodservice operators who have implemented a 100% self-delivery program have noted concerns and now use a **hybrid delivery program**.

Key Term

Hybrid delivery program: An arrangement in which a foodservice operator and a third-party delivery app jointly offer delivery services for the operator's store(s).

Operators of a hybrid delivery program deliver a portion of their off-premises guest orders, while a third-party delivery app is utilized for the balance of the orders. This hybrid model is gaining more attention in the industry as operators seek to manage costs while still supplying their guests with high-quality delivery services. Properly established hybrid models of delivery can allow operators to:

1) Manage delivery times

Utilizing a hybrid approach, operators can select specific days or hours when they prefer to do their own self-delivery. For example, an operator may choose to deliver orders only during lunch and dinner hours or just on weekends. At all other (slower) times, orders would be delivered by third-party delivery partners.

2) Manage delivery distance

Utilizing this strategy, operators would select to deliver orders only within a specified delivery radius. For example, if an operator set a range of two miles, a menu order that is to be delivered at a greater distance will automatically be outsourced to the third-party delivery partner.

3) Manage order value

As earlier addressed, third-party delivery companies charge commissions based on an order's total value. Therefore, operators could elect to self-deliver only orders that exceeded a selected amount. For example, an operation may accept for in-house delivery only those orders over $100. All orders under $100 would be delivered by the third-party delivery partner.

4) Manage order quantities

A foodservice operation might have a fixed number of drivers, and the operator could establish a limit to the number of deliveries allowed per driver, For example, if an operator set a limit of two delivery orders per driver, and the operator had three drivers on the work schedule, the 7th order would be outsourced.

5) Manage driver availability

Utilizing this strategy, operators would elect to receive orders only when their own drivers were available. If all the operator's drivers were busy, the order would automatically be outsourced to the third-party delivery partner.

6) Manage delivery parameters

Using this strategy, foodservice operators could establish specific delivery zones. For example, an operator could choose to self-deliver orders valued at $20 or more only within a one mile radius, but $100 or more orders could be delivered up to five miles.

Technology at Work

Given the controversy surrounding third-party delivery partners, it is likely that the future will yield even more organizations developing tools to assist operators as they navigate this challenging area of guest service.

One example of this is the Cartwheel company (www.cartwheel.com). Their state-of-the-art delivery software can be used by those operators electing self-delivery, 100% outsourced delivery, or a hybrid solution that combines the best features of both approaches.

Increasingly, foodservice operators will seek to provide delivery services to their guests using the most cost-effective procedures . To stay updated about companies offering delivery options to foodservice operators, enter "meal delivery assistance for restaurants" in your favorite browser and view the results.

This chapter has addressed how foodservice operators can increase their online presence by developing local links and by monitoring relevant information available on third-party operated sites. It also explored the advantages and disadvantages of partnering with external companies operating third-party delivery apps.

Third-party-operated websites of many types should be an important part of a foodservice operation's online marketing efforts. Those third-party-operated websites that focus primarily on content posted by the site's *users*, rather than the site's operators, are an important category of third-party sites, and they should always be carefully monitored. How this can be done promptly and professionally will be addressed in detail in the next chapter.

What Would You Do? 10.2

"So what did the salesperson say?" asked Amir, the assistant manager of the Campus Deli.

Amir was talking to Peggy Richards, the owner of the Campus Deli, which is a popular New York–style delicatessen located near a major college campus. Peggy's restaurant featured traditional deli fare like corned beef and pastrami sandwiches, homestyle soups, and traditional cheesecakes. Peggy employed a small number of delivery employees who delivered call-in and online orders to students and faculty living in or near the college. Business was good, and orders placed for delivery were increasing significantly.

"She said that, with our increasing demand for off-site delivery, it is a good idea to partner with them," replied Peggy.

"That makes sense to me," said Amir, "what do you think?"

"I don't know. I think we should keep our own drivers and work to decrease demand for off-site delivery. I think we can do that by offering more margin-friendly curbside and takeout options to our guests. If we reduce the prices for guests coming to get their own food, we'll be moving the amount we would have paid to the delivery company back to our own guests. I think I'd like to try that approach first."

Assume you were Peggy. What would be some risks associated with seeking to de-emphasize the off-site delivery of guest orders? What could be potential advantages of implementing such a strategy?

Key Terms

Link

Local link

Convention and Visitors Bureau (CVB)

Income and expense statement

Uniform System of Accounts for Restaurants (USAR)

Net income

Merchant services provider

Cash flow

Hybrid delivery program

Operator's 10-Point Tactics for Success Checklist

Evaluate your need for, and the current status of, each of the following operational tactics. For those tactics you think are important, but not yet in place, develop an action plan for its implementation including who will be responsible for the tactic's completion and the target date by which it should be completed.

Tactic	Don't Agree (Not Done)	Agree (Done)	Agree (Not Done)	If Not Done	
				Who Is Responsible?	Target Completion Date
1) Operator understands the importance of local links in the expansion of their online marketing presence.	____	____	____		
2) Operator has established links with appropriate websites in their local community.	____	____	____		
3) Operator has conducted a Google search of their own operation to review the resulting SERP.	____	____	____		

(Continued)

Tactic	Don't Agree (Not Done)	Agree (Done)	Agree (Not Done)	If Not Done	
				Who Is Responsible?	Target Completion Date
4) Operator has identified those third-party-operated websites that are listed highest on the SERPs for their operation.	____	____	____		
5) Operator has established relationships with those third-party-operated websites that are listed highest on the SERPs for their operation.	____	____	____		
6) Operator has included the maintenance of professional relationships with those third-party-operated websites listed highest on the SERPs for their operation in their next period marketing plans.	____	____	____		
7) Operator has assessed the value of providing off-site delivery services for guests of their own operation.	____	____	____		
8) Operator has carefully reviewed the advantages of partnering with third-party delivery apps when providing off-site delivery services for guests of their operation.	____	____	____		
9) Operator has carefully reviewed the disadvantages of partnering with third-party delivery apps when providing off-site delivery services for guests of their operation.	____	____	____		
10) Operator has assessed the costs and benefits of implementing their own off-site delivery services for guests of their operation.	____	____	____		

11

Marketing Management on User-Generated Content Sites

What You Will Learn

1) The Popularity of User-Generated Content (UGC) Sites
2) The Importance of Advertising on User-Generated Content (UGC) Sites
3) The Popularity of User-Generated Content (UGC) Sites Featuring Guest Reviews
4) How to Positively Impact Scores Posted On User-Generated Content (UGC) Review Sites
5) How to Respond to Negative Reviews Posted on User-Generated Content (UGC) Review Sites

---

Operator's Brief

In this chapter, you will learn that user-generated content (UGC) sites are among the most popular sites on the Internet. Therefore, operators should carefully consider the advantages of monitoring and advertising on the most popular UGC sites. They should be especially interested in the UGC sites that feature guest reviews of foodservice operations the writers have visited.

Those who view these social media sites are often heavily influenced by the reviews they read. New customers reading many positive online reviews will likely feel they can trust the operation to provide them with a high-quality meal experience. Alternatively, large numbers of negative reviews tend to discourage potential customers from choosing the operation.

Most foodservice operations want to increase the total number of reviews posted online, and there are reasons for this interest. These include the impact on the search engine results page (SERP) generated when online users seek information about an operation and the likely increase in the operation's future guest counts.

(Continued)

> Sometimes the ratings posted online about an operation are below average. Then it is important that operators take decisive steps to remedy the situation(s) causing the poor reviews. Foodservice operators should also promptly and professionally respond to any negative user reviews that have been posted. Doing so increases an operation's listing on a SERP, and it helps reassure future customers that problems stated in the online reviews have been properly addressed and corrected.

CHAPTER OUTLINE

The Importance of User-Generated Content (UGC) Sites
Advertising on User-Generated Content (UGC) Sites
Popularity of User-Generated Content (UGC) Sites Featuring Guest Reviews'
    Historical Perspective
    Motivation of Reviewers
Improving Scores on User-Generated Content (UGC) Review Sites
    Increasing the Number of User Reviews
    Increasing the Scores on User Reviews
Responding to Negative User Reviews

The Importance of User-Generated Content (UGC) Sites

In Chapter 4, a user-generated content (UGC) site was described as any website in which most content including images, videos, text, and audio have been posted online by the site's visitors. There are numerous social media sites (see Chapter 9) such as Instagram, Twitter, Facebook, Vine, Pinterest, TikTok, Snapchat, and YouTube. Note: The importance of enhancing an operation's presence by creating and managing its own accounts on these sites was addressed in the previous chapter.

To best understand the ever-increasing popularity of UGC sites, operators should recognize, that in the early days of Internet usage, users logged online, selected a website to view, and then read or watched the content that was posted there. In most cases, they saw content that had been professionally developed and then posted online by a business or organization. The operators of the website had complete control over their sites' content.

This arrangement was a good way to provide a site's viewers with information about a business. However, it was not a good way for site users to communicate with the business or with other users. This makes sense because when an operation or brand posts information about itself online, it is possible to ensure the information is accurate. In effect, then, these early methods were simply a high-tech form of one-way communication (see Chapter 4), from website operator to viewer.

As previously addressed, the best communication systems are two-way systems that allow the exchange and sharing of information. As social media sites emphasizing shared content continued to develop, they filled a desire by Internet users for information sharing, rather than just information gathering.

Today the most read and viewed content on the Internet is that contributed by visitors to UGC sites. The amount of information posted online by users is extraordinary. On any given day, it is estimated that 350 million photos are uploaded to Facebook, 95 million photos and videos are shared on Instagram, and over 500 million tweets are posted on Twitter.

Professional marketers recognize the power of UGC sites because they know that people more readily trust the recommendation of a real person over what a business says about itself. In a recent study, 56% of consumers said they more readily trust user-generated content and that it was what they most wanted to see from businesses. Note: Only 15% of the consumers said content created by businesses was what they wanted to see.[1]

Since the popularity of specific social media sites will likely continue to rise and fall and new sites will be introduced, their importance to foodservice operators as they market their businesses cannot be overstated. This is true both in terms of their use as an advertising channel, and their importance related to the guest reviews and ratings posted on these sites.

Advertising on User-Generated Content (UGC) Sites

UGC sites are popular, and they can be an excellent advertising channel (see Chapter 8) for many foodservice operators. It is not usually necessary for a foodservice operation to have an active social media account on a UGC site to advertise on it. So, for example, a foodservice operator may advertise on Facebook or LinkedIn without creating and maintaining an active Facebook page or LinkedIn account.

Technology at Work

The specific ways that social media sites permit foodservice operators to advertise varies based on the site. The specific form of an ad to be posted on a social media site is also most often best determined based on the site itself. For example, sites featuring video ads such as YouTube and TikTok typically are the most effective because visitors to those sites expect to see videos. In contrast, other sites print advertisements that may (or may not) include photographs.

(Continued)

---

1 https://www.nosto.com/blog/what-is-user-generated-content/ retrieved September 1, 2022.

Navigating all these social media sites and their varied advertising forms, requirements, and benefits can be overwhelming for some foodservice operators. This is also true about the effectiveness of various metrics site operators provide to their advertisers. Fortunately, there are numerous companies to assist advertisers as they create and post effective social media ads.

To review some of these companies and the services they provide, enter "social media advertising assistance for restaurants" in your favorite browser and view the results.

For foodservice operators, social media advertising can be viewed as a tool that can create and maintain an effective online presence. Advertising on social media sites is often referred to by marketing professionals as **paid social** advertising.

Each social media site will have its own method of charging for the advertisements it displays. In some cases, the fee is a flat dollar amount based on an ad's size and the amount of time the ad is displayed. In others, the fee is based on the number of times the advertisement is accessed (clicked on) by social media users.

Key Term

Paid social (advertising): The use of ads on social media sites for which an operator must pay a fee.

Operators should remember that, when promoting their own proprietary Internet sites, the emphasis should address target keywords and terms (see Chapter 9). In contrast, on social media sites, it is more important to target specific audiences, interests, and behaviors.

For example, some social media sites attract younger audiences, while other sites attract audiences interested in specific content formats irrespective of their age. Regardless of the social media sites on which they prefer to advertise, foodservice operators can employ several key actions as they plan and implement their paid social efforts:

1) Define the target audience

In most cases, UGC sites attract specific types of audiences, and operators can develop their target audience based on factors including location, job title, gender, hobbies, and relationship status. Remember that the more tightly defined an audience is, the more specifically ads can be tailored to these potential customers.

2) Create memorable content

Foodservice operators choosing to utilize paid social advertising compete with a tremendous amount of other information and advertisements placed on social media sites. Therefore, posted content must be interesting, engaging, and eye-catching to be competitive. Also, it is important to remember that the attention span of many social media site viewers is limited. Effective ads must immediately grab the viewer's attention, be memorable, and be kept short so they can be viewed quickly.

3) Create ad variations

Those who manage social media sites can provide foodservice operators with excellent metrics regarding an ad's performance. The number of clicks, number of views, and time spent viewing an ad are all common metrics carefully collected by social media site operators. Slight changes in wording, photographs used, and ad size can create significantly different results. When operators create ad variations, they can continue ads that perform best on the social media site.

To best track their effectiveness, some ads should have "**call to action**" features. Examples include instructions to:

- Call us now for more information
- Click here to make a reservation
- Sign up for our loyalty program
- Purchase a gift card
- Click here to go to our website
- Add your e-mail address to our mailing list
- Watch this exciting video

Key Term

Call to action: A piece of content intended to induce a viewer, reader, or listener to perform a specific act, typically taking the form of an instruction or directive (e.g., *buy now* or *click here*).

4) Change ads frequently

Most users who visit their social media accounts do so frequently (in some cases, several times per day). Therefore, ads that remain unchanged for long time periods will likely be viewed once and then skipped over. The best advice: keep the interest and engagement of a social media site's users high by changing ads frequently.

5) Analyze the data produced

Operators of social media sites can provide significant amounts of an ad's effectiveness-related data. This should be regularly and carefully reviewed and analyzed to determine if the return on investment (see Chapter 12) for the advertisement is positive.

Popularity of User-Generated Content (UGC) Sites Featuring Guest Reviews

Many UGC sites enable the site's users to post reviews of restaurants they have visited, and some UGC sites significantly feature restaurant reviews or are even devoted almost exclusively to these reviews.

• Yelp	• Google Business Profile
• OpenTable	• Facebook
• TripAdvisor	• Foursquare
• Zagat	• Third-party delivery app sites
• Zamato (formerly Urbanspoon)	• Reddit

Figure 11.1   Currently popular UGC review sites

While the popularity of any given UGC review site will likely increase or decrease over time, Figure 11.1 lists 10 of the most popular UGC review sites currently in operation. Note especially that Figure 11.1 includes third-party delivery app sites (see Chapter 10) such as Grubhub, DoorDash, and Uber Eats, and nearly all of these delivery services offer a user-review option.

To better understand the importance to foodservice operators of UGC sites featuring guest-contributed restaurant reviews, a 2022 survey found:

✓ 99% of consumers have used the Internet to find information about a local business in the last year.
✓ More consumers are reading online reviews than ever before. In 2021, 77% "always" or "regularly" read them when browsing for local businesses including foodservices (up from 60% in 2020).
✓ The percentage of people "never" reading reviews when browsing local businesses has fallen from 13% in 2020 to just 2% in 2021.
✓ 89% of consumers are "highly" or "fairly" likely to use a business that responds to all online reviews.
✓ 57% say they would be "not very" or "not at all" likely to use a business that does not respond to reviews.
✓ In 2021, just 3% said they would consider using a business with an average star rating of two or fewer stars, down from 14% in 2020.[2]

Historical Perspective

Hospitality professionals reviewing bestselling foodservice marketing books published in the 2010s and earlier would find virtually no information about the importance to marketing of UGC sites displaying restaurant reviews. However, times change! Today online UGC sites featuring restaurant reviews are extremely popular, and their impact on marketing foodservice operations is well-established and highly significant.

Published customer reviews of restaurants are not new. Beginning in the late 1800s, the *New York Times* regularly published restaurant reviews written by food critics employed by the newspaper, and that newspaper still does so.

---

2 https://www.brightlocal.com/research/local-consumer-review-survey/ retrieved August 30, 2022.

The French-based Michelin Tire Company published its first hotel and restaurant guide in 1900 and awarded its first coveted "stars" in the 1926 edition. In 2005, Michelin published its first American guide covering 500 restaurants in the five boroughs of the New York City. To produce its guides, Michelin recruited a team of anonymous inspectors to visit and review the restaurants.

Find Out More

In the 1960s, the French chef Paul Bocuse, one of the pioneers of what would later become known as "nouvelle cuisine," famously stated, "Michelin is the only guide that counts."

While today there are many sources of restaurant reviews that "count," it is generally accepted by foodservice professionals that earning one or more Michelin stars is a crowning achievement for chefs and their operations.

In the Michelin ratings:

One star describes "A very good restaurant in its category."
Two stars describes an operation that has "Excellent cooking, worth a detour."
Three stars is an operation that provides "Exceptional cuisine, worth a special journey."

The difficulty of earning three stars is illustrated by the fact that, as of 2020, there were only 14 Michelin-rated three-star restaurants in the United States and only 135 of these properties in the entire world.

While most foodservice operators are unlikely to earn a Michelin star, operators can boost their chances of earning a star (or simply being very highly rated by their guests!) by paying attention to every detail of their foodservice offering. Although food taste, attractiveness, and attention to details are important, it is the entire experience from selecting where, what, and how to order that is most important to guests.

Successful operators recognize that, in today's busy economy, some foodservice customers expect a meal that offers a value—not a comprehensive dining experience. Effective ads that focus on guests' needs and the "best" ways to deliver them will remain an important ingredient in the "recipe" for success.

To learn more about the factors Michelin uses to rate restaurants, and how you can apply some of this information to your own operation, enter "how Michelin stars are earned" in your favorite browser and view the results.

In 1979, Tim and Nina Zagat began the "Zagat Survey." The philosophy behind the Zagat Survey was unique in that, instead of reading one review written by a paid restaurant critic, the public would be better served by a rating based on hundreds of guest responses.

By compiling data from surveys of actual customers, the Zagat Survey rated restaurants on a 30-point scale in the categories of food, décor, service level, and cost. In 2016, Zagat (then owned by Google) switched to a system that scores restaurants on a scale of one to five stars. However, the idea of actual users (rather than only professional reviewers) sharing their dining experience thoughts proved to be extremely popular. Today a variation of this Zagat approach is still used, and many popular UGC sites allow users to post their restaurant review comments directly on the site.

Motivation of Reviewers

Some foodservice operators are wary of reviews guests may post on social media and UGC review sites, and a cautious reaction is not surprising. In many cases, those who operate foodservice facilities deal directly with guests when a problem arises. If, for example, an operator serves 100 guests in one day, and five customers that day had a complaint, the operator might feel that "many" customers are dissatisfied. In this example, however, 95% of this operator's customers did not encounter a problem worth mentioning.

One should recognize that reviewers are more motivated to post *positive* reviews about foodservice operations rather than negative reviews. While consumers do share experiences when they are dissatisfied, they are much *more likely* to share experiences that show their selection of a foodservice operation was a good one.

Online marketing research shows that younger consumers are more likely to post reviews than older consumers.[3] Regardless of age, however, foodservice guests typically feel compelled to submit ratings and reviews if they:

✓ Have a positive experience
✓ Are genuinely wanting to help the operation improve its products or services
✓ Are motivated to give a special "thanks" to one or more individual staff members
✓ Have been incentivized in some way to do so
✓ Have a negative experience
✓ Want to contribute to the social media community of which they are a member
✓ Want to encourage (warn) other site users about an operator

Recognizing the importance of ratings and user reviews to their online reputations, some operators worry about the possibility of competitors writing negative reviews to harm their business. Certainly, this can occur when, for example, a competitor (or disgruntled former employee) feels threatened and makes online statements with the goal of damaging another business's reputation.

---

3 https://reviewmonitoring.com/demographic-profile-of-the-average-online-review-writer/ retrieved September 29, 2022.

If an operator suspects that another business is responsible for a negative review posted online, the operator should not respond in anger. Rather, a professional response like the following should be posted:

> *Dear . . ., thank you so much for your feedback,*
>
> *Unfortunately, we have searched and can find no records or recollections that indicate you were a customer in our operation. If you would please provide more details such as a receipt showing the day and amount purchased from us, we will be happy to find a solution to the issues you encountered.*
>
> *We currently have no record of your name or a visit that matches your description of what you encountered. I sincerely hope you will please send me an e-mail or give me a call so I can personally address this immediately. Again, thank you for your input.*
>
> *Best wishes*
> *Manager*

A professional response such as the above is helpful in two ways. First, one cannot address a problem if he or she is not aware of it. Second, the operator may be incorrect, and the actions noted in the review can be authentic and genuine. Then the alert provides the operator with an opportunity to resolve the issue and, hopefully, receive an updated and more positive review from the customer.

If, however, it is a **fake review** the operator is demonstrating that, as a business, it takes responsibility for issues that arise. It tells site readers that this may be a fake review and, when other site visitors read it, they can note that the review writer may never have visited the operation or purchased its products and services. Another potential benefit: some site readers may make additional visits to an operation's proprietary website to note any follow-up issues that occur. In effect, this initial negative review can be resolved from the perspectives of both the review writer and the public website visitors.

Key Term

Fake review: A positive, neutral, or negative review that is not an actual consumer's honest and impartial opinion or that does not reflect a consumer's genuine experience with a product, service, or business.

Fake reviews can come from several sources including:

✓ Business operators who write their own positive fake reviews to attract new customers
✓ Former employees who may post negative reviews because they have been laid off or involuntarily terminated
✓ Current employees, friends, or family members who write positive reviews to help the business increase its overall ratings and scores
✓ Customers who write unfair negative reviews intended to help them receive a refund or other benefit

---

Find Out More

UGC sites featuring guest reviewers will not remove a negative review just because an operator does not like it. Nearly all social media sites, however, offer businesses the opportunity to challenge reviews they believe to be inaccurate.

The procedures used in each site are similar, and the specific steps to be taken are generally provided under the "Support" section of the site. In most cases, the site will require the operator to provide as much information as possible about the alleged "fake" review as the site's operator seeks to verify its authenticity.

To learn more about how operators can remove fake or unfair reviews enter "how to challenge and remove a fake online review" in your favorite browser and view the results.

Marketing research has shown that the majority of reviewers want to help the business by leaving a favorable review. If a foodservice guest has a positive experience with an operation, they want to let others know about it so the readers of the review can have the same enjoyable experience. Reviewers also believe they are rewarding the company for the excellent food or service they received.

When negative reviews are written it is likely that a guest had one or more negative moments of truths (see Chapter 5) that were not corrected on-site or upon (or after) a product delivery. This demonstrates that problems solved immediately lead to higher online scores because guests with corrected negative experiences are less likely to write about the problem if it was resolved.

What Would You Do? 11.1

"We've got to fix this," said Clara, the dining room manager at Butterflies Bistro.

"Fix what?" asked Elijah, the Bistro's kitchen manager.

"This average score we're getting on the 'DineInTown' website," replied Clara.

The "DineInTown" website featured guest reviews of local restaurants, and it was a popular site with many postings. The site allowed reviewers to rank restaurants on a scale of 1–5. (1 being a low score, and 5 being the highest score.)

"We are averaging a score of 3.8," said Clara. "I do not expect us to get all 5s, but we should be at least in the 4-plus range. The interesting thing is that we tend to get almost all 5s because reviewers say they love our food, and reviewers who give us a score of 1 almost always refer to slow service. These are mostly guests who visited us on a Friday or Saturday night. With lots of 5s and then a few 1s mixed in, we end up with a 3.8 average. That's what we need to fix."

Assume you were Clara. What do you think your guest reviews are telling this operation about its staffing patterns on Friday and Saturday nights? How important would it be for you carefully manage reservation capacity and service and production staffing levels on these two busy nights?

Improving Scores on User-Generated Content (UGC) Review Sites

The scores or ratings a foodservice operation receives on UGC sites are extremely important. High scores are desirable, and the online reviews posted by actual customers make it very convenient for potential customers to obtain in-depth information from others with little effort. As a result, UGC websites such as Yelp and TripAdvisor offer reviewer content that is often considered more dependable and trustworthy than business advertisements.

As addressed earlier in this chapter, the motivation for making and posting foodservice operation reviews vary, and there are key reasons why those who search for high-quality UGC sites featuring reviews do so. These include:

1) Reducing risk

    Potential customers reading business reviews want to reduce their risk of making a mistake. When, for example, reviewers consistently state that service is poor or food quality is below standards, potential customers feel they can use that information to avoid disappointment from making a poor decision to visit a foodservice operation.

2) Minimizing search time

    In nearly all communities foodservice customers can select from a wide range of foodservice operations. High-quality UGC sites feature reviews that enable potential customers to learn about poor scores easily and quickly and with minimal search times.

3) Minimizing buyer's remorse

    A consumer may experience **buyer's remorse** after buying something for several reasons. These include paying too much, buying a product that did not meet expected quality levels, and receiving especially poor service during the purchase experience. In these and other cases, a buyer's remorse usually stems from anxiety that the decision to buy was a poor one.

Key Term

Buyer's remorse: A feeling of regret that occurs after buying a product or service.

In many cases, a consumer can overcome buyer's remorse by returning an ill-advised purchase to a store. Consider a buyer who makes an impulse decision to buy an expensive laptop computer. The next day the person decides that the purchase decision was unwise. In this scenario, the laptop can, in most cases, be returned to the store for a full refund.

When making purchases from a foodservice operation, however, the purchase decision cannot be reversed. After the product is consumed and the services are rendered, the purchase is irrevocable. As a result, more consumers are concerned about minimizing buyer's remorse when making choices about food establishments than when making choices about other products from other businesses.

4) Experiencing groupthink

"Groupthink" is a term commonly used to describe a belief in the wisdom of group members' feelings and opinions. In the case of UGC sites featuring foodservice reviews, potential customers often believe that, if many others who have posted reviews believe an operation is "good," the group's opinion is likely correct. Similarly, if a review site suggests an operation is "poor," then possible customers will likely find the operation to be poor as well.

Groupthink can also impact foodservice review sites in another way. Market research consistently shows that site visitors pay more attention to, and have greater confidence in, scores based on a larger (not smaller) number of reviews.

In nearly all cases, foodservice operators should implement procedures that yield large numbers of positive reviews on the most popular UGC review sites.

Increasing the Number of User Reviews

Online guest reviews are today's equivalent of "yesterday's" word-of-mouth advertising. Traditionally, those desiring information about a foodservice operation asked friends or family members for recommendations. Today, Internet users turn to online customer reviews to help determine the operations they will visit for the first time.

For the majority of UGC sites featuring customer reviews, three factors determine how high an operation will be listed on its search engine results page (SERP) (see Chapter 9). These are:

✓ Number of reviews (more reviews are better than fewer reviews)
✓ Quality (scores) of reviews (good reviews are better than poor reviews)
✓ Recency of reviews (recent reviews are better than older reviews)

Foodservice operators can employ several strategies to increase the number of reviews their operation receives:

1) Featuring positive reviews

   A foodservice operation's proprietary website was described earlier as an operation's single most important online communication tool (see Chapter 9). Therefore, featuring positive reviews on the home page of an operation's proprietary website is always a good idea. These reviews can be posted in their entirety, or they can be highlighted with the use of a specific "Reviews" tab that, when clicked on, displays positive reviews of the operation. This approach encourages others to post additional reviews because they see that reviews are valued and showcased by the operation.

2) Linking to popular social media sites

   When foodservice operators have created their own accounts on popular social media sites, those sites should be linked to the operators' proprietary websites. For example, it should be easy for a website viewer to click on a Facebook link placed on the proprietary site and immediately be taken to the operation's Facebook page where reviews can be posted. Generally, foodservice operators should create accounts on the social media sites their guests are most likely to utilize.

3) Asking current customers to leave a review

   In many cases, foodservice personnel can simply ask guests at the end of their meal if they would be willing to leave a review. This can be done in a nonintimidating way by servers who simply say, "If you enjoyed your meal with us today, we'd love for you to leave us a review!" In many cases, this will result in guests leaving a positive review on the social media UGC sites they themselves use. For pickup and third-party delivery orders, a note from management can be placed within the food packaging requesting that the guest leave a review and providing easy instructions on how to do so.

4) Following up with recent customers

   Those foodservice operators collecting e-mail addresses or phone numbers from the guests who have visited their operations can send a followup e-mail or text simply asking the guests if they would be willing to post a review.

   The e-mail or text should include directions or direct links to make writing and posting the reviews very easy. When making an "ask" for a review, operators should seek to:

   - Make sure the e-mail or text comes from a real person (i.e., Bobjones@bobspizza.com, rather than support@bobspizza.com).
   - Write the e-mail as a personal request from the sender using the first name of the recipient, if possible
   - Include a button that links directly to the review site
   - Make the request as short as possible to minimize reading and response time

5) Rewarding those who leave a review

One good way to reward those contributing a review is by offering them something such as a 10% off coupon for future purchases. This strategy can be employed even if the discount is not disclosed to the guest before the review is written. In that scenario, it is highly likely the guest will be delighted when they receive the unexpected discount and may even tell others about it (which may encourage even more positive reviews!).

6) Making it easy

Perhaps the most important thing foodservice operators can do to increase their number of reviews is to make it easy to post a review. To do this, it is important for foodservice operators to understand the review posting process. To illustrate, consider the process used to post a Google review.

Google is an extremely popular UGC site that features restaurant reviews following several steps:

1) Open Google maps
2) Search for the operation's name
3) Click on the operation's name to pull up its Google business profile
4) Scroll down to the review section
5) Click on "Write a review"

Let's consider the situation in which a guest who just finished a meal wants to leave and says to an operation's manager, "My meal was wonderful! How can I leave a review?" The manager's response in this scenario could be:

> Thanks very much for asking about the review process, and you can leave a Google review. It's easy. Just go to Google Maps and do a search for us by entering our name in the search bar. When you pull up our listing, scroll down the page until you come to the review section. Then just click that button, and you can write a review.

Or the response could be:

> Thanks. Just go to our website and click on the Google Review link button located on our homepage!

Clearly, the manager using the second response is more likely to get the review! Foodservice operators should provide direct links to review sites that are important to them as well to those that are currently of most importance to their target markets.

Increasing the Scores on User Reviews

The score given on a UGC site featuring guest reviews allows the reviewer to rank the quality of the experience that occurred when the reviewer interacted with the operation. When operations have met or exceeded their expectations, they are highly ranked, and lower rankings are given when the reviewer's expectations were not met.

Various UGC sites featuring user reviews have different scoring systems. For example, on Yelp, scores can range between 1 and 4 "Stars." On Google and Facebook, the scores range between 1 and 5 "Stars." On TripAdvisor, reviewers assign a score of 1 to 5 that results in the display of filled or partially filled "Circles."

Grubhub uses a 5 "star" rating system. It also publishes an additional summary, "Here's what people are saying," about a business. For example, the review for a single operation might say, "Here's what people are saying: 93% Food was good, 88% Delivery was on time, 82% Order was correct."

On Google and TripAdvisor, among others, the sites' operators also use algorithms to rank operations. For example, a search for "Best pizza near me" will produce a SERP that ranks pizza restaurants "in order." However, the ranking includes more factors than just the average reviewer's score because frequency and recency of reviews are also factored into the review.

While the specific scoring and ranking systems used by UGC review site operators vary, the most essential fact is that foodservice operators must recognize that the scores matter. Online market research indicates that 70% or more of site visitors reviewing online ratings only continue to read information about an operation if its ratings are within the top two scores (3 and 4 for a four-point rating system, and 4 and 5 for a five-point rating system). Therefore, scores achieved by an operation directly affect the amount of business they will do. In fact, a comprehensive study by Professor Michael Luca at the Harvard Business School found that a 0.5-star increase in average Yelp rating generates a 9% revenue increase in revenue for an independent restaurant.[4]

Since UGC review scores do matter, proactive operators can employ several strategies to increase the average review score their operations receive:

1) Monitoring the reviews

Foodservice operators monitoring their online reviews can learn what their customers think about them. Certainly, online reviews communicate to potential guests who read them, but they also provide information to foodservice operators. Many foodservice operators conduct formal guest surveys to learn what their guests think about them but, in most cases, guest reviews posted online provide the same information.

---

4 https://dl.acm.org/doi/pdf/10.1145/3432953 retrieved August 30, 2022.

Often, foodservice operators can gain insight into their own businesses by reading their reviews. For example, assume that a consistently high number of reviewers indicate that parking when arriving to pick up their orders is a problem for guests. In this example, the operation is clearly being told that it needs to take steps to improve its product pickup system. This may entail, for example, designating special parking spaces for pickup orders, or providing servers who can deliver orders to guests' arriving vehicles. Similarly, if review scores posted on third-party delivery app sites are low, it may indicate a problem with third-party delivery order preparation, packaging, or delivery to drivers.

Another key reason for monitoring reviews is to measure an operation's success in improving its scores. To illustrate, assume that a foodservice operation was achieving a score of 4.2 stars on a popular user review site. Assume further that the operator wanted to increase that scoring average to 4.5 in the coming marketing period. In this example, the monitoring of score movement would be important for the operator to determine whether its marketing strategy and tactics in this area are meeting with success.

2) Encouraging reviews from satisfied customers

SERP results are weighted based on the number of reviews and the recency of reviews, but they are most heavily influenced by the actual scores given by reviewers who submit a rating.

UGC sites featuring user reviews use an **arithmetic average** to calculate the average rating of a business. The more positive reviews received, the higher the average rating will be. It is always helpful to encourage reviews from guests who are highly satisfied with their experience. This can often be determined by asking servers to confirm a guest's experience was positive and then follow up by asking for a review. "Regular" guests who visit, call, and/or order via the internet could also be asked to complete a survey.

It is also a good idea to encourage reviews from an operation's customers who, by their status of being a "regular," have indicated they are highly pleased with the food quality and service they always receive.

Key Term

Arithmetic average:
The simplest and most widely used measure of a mean (average); calculated by dividing the sum of a group of numbers by the count of the numbers used in the series. For example, if the reviewers' scores on a review site are 3, 4, 4, 5, and 5, and the sum is 21, the arithmetic mean is divided by five and equals 4.2 (21/5 = 4.2).

3) Fixing problems when they occur

An arithmetic mean is greatly reduced when review scores are low, so low review scores should be avoided, when possible. The best way to do this is by ensuring that any problems that occur during a guest's visit are immediately resolved *before* the guest leaves the operation.

Problems in food quality or service delivery can occur in any foodservice operation. When operators proactively inquire about guests' experiences, those with a negative moment of truth (see Chapter 5) can often have their problems immediately corrected. Doing so greatly reduces the chances that the guest will post a negative online review and may even result in a positive review describing how the operation corrected its mistake.

4) Responding to reviews

Recent reviews *and* recent responses weigh heavily into SERP results. A foodservice operator receiving dozens (or even hundreds!) of reviews need not respond to every review, but the importance of responding to all negative reviews is addressed later in this chapter.

Operators should regularly (several times weekly, if possible) respond to online reviews. Creating a dialogue between the operator and customers is perceived positively by those reading online reviews. Responses to reviewers also indicate to site visitors that the business considers and values the quality of its customers' experiences as the operation is managed.

The responses to a positive guest review need not be long and detailed. A simple "thank you" and an expression of pleasure that the guests received a positive experience typically suffice.

5) Providing excellence in food quality and service

Guests selecting a foodservice operation anticipate they will have a good experience, and they are typically predisposed to think their decision to visit an operation was a good one. Providing excellence in food quality and service is the best way to verify that the decision was, in fact, a good one.

An operation that works hard to provide high-quality guest experiences is likely to avoid poor reviews. Paying close attention to food and service details encourage high review scores because doing so results in satisfied customers. These review scores will also "invite" potential guests reading high-ranking online reviews to visit in the future.

Responding to Negative User Reviews

Even the best-managed foodservice operations will encounter some negative reviews posted on UGC review sites, and operators should respond to them promptly and professionally for several reasons:

1) A boost in ranking or listing: Many data integration programs are developed to respond to, and reward, business-to-consumer interactions. When these programs detect that a business is regularly communicating with its customers that business will be placed higher on a SERP page.

2) Showing the business cares: When customers complain, they typically want to know that someone hears them. A response (of almost any kind!)

demonstrates that management does care about its guests' challenges and is interested in addressing them.

3) Minimizing the damage: Once a legitimate negative review has been posted online, it cannot, in most cases, be taken down. However, when readers or viewers of a negative review also immediately encounter a sincere and timely professional management response, the reputational damage done by the negative review is lessened.

4) Demonstrating professionalism: A rapid response to negative reviews shows that a business is concerned about its customers, is responsive to them, and will respond quickly and professionally when issues arise. These are important traits all consumers look for when they buy products or services.

Foodservice operators do not have the luxury of being able to ignore negative reviews in the hopes they will go away because, in nearly all cases, they will not do so. Ignoring negative reviews increases their impact because a reply is not just a response to that person. Instead, it is a message to all others who will read or see the review including potential guests.

A lack of response by a business tends to confirm the general position of the reviewer (i.e. this business does not care about its customers!). Figure 11.2 lists guidelines for how foodservice operators can positively respond to negative online postings about their business.

When responding to negative (or positive!) reviews, operators must always remember that their responses should be well-written and grammatically correct.

---

1) Thank the reviewers for taking their time to give feedback. (It is always best to start a response in a positive way.)
2) Apologize if the customer is correct about what happened.
3) Avoid arguing if the customer is incorrect about what happened. Instead, apologize there was a misunderstanding.
4) Provide a very brief, but clear and direct explanation of what caused the problem.
5) Assure the site's readers that specific actions have been taken to avoid a repeat of the problem.
6) Offer a direct line of communication between the business and the negative reviewer (via e-mail, text, or phone) to receive a personalized apology.
7) Conclude the response by directly quoting any possible part of the reviewer's comments that were positive (for example, if the reviewer says the server was "friendly" but the soup was "cold," mention that it was good to hear the reviewer found the server to be "friendly."
8) End the posting by *again* thanking the reviewer for feedback and helping to make the operation even better because of it.
9) Invite the customer to give the operation another chance to provide excellent service and products.

Figure 11.2 Guidelines for Responding to Negative Online Reviews

Technology at Work

Foodservice professionals in all segments of the hospitality industry understand that the world is more connected than ever due to the wide reach of the Internet. Online reviews are tools consumers use to communicate with each other about the quality of the foodservice operations they have visited. While most reviews posted online will be good ones, negative comments about a business or brand can create significant damage. Online review monitoring companies can offer significant value to foodservice operators of all sizes, and they have programs to continually scan the Internet in search of positive and negative reviews about their clients.

Sometimes consumers post unfair or untrue reviews. If this occurs, an online review monitoring company can often help mitigate them. This will protect an operation's reputation because the company understands the required methods to help operators remove fake reviews from a site. They can also assist in identifying positive reviews and disseminating them as widely as possible.

To examine some of the online review monitoring companies and to learn about the services they can provide, enter "assistance in monitoring online reviews" in your favorite browser and view the results.

This chapter addressed the importance of UGC sites and especially the significant role of those sites that feature guest reviews. The focus of the next chapter is on how foodservice operators can measure and evaluate the effectiveness of their current marketing efforts so they might be improved in the future.

What Would You Do? 11.2

"I'm already spending a ton of time monitoring the sites and responding to the few negative reviews we receive," said Roz Jaffer, the assistant manager of Zaytoons' Mediterranean restaurant.

Zaytoons' was a popular operation featuring traditional Mediterranean dishes including Chicken Shawarma, Lamb Kabobs, Shish Kafta, and Kibbi. The previous manager of the restaurant had assigned Roz the task of monitoring UGC sites featuring customer reviews and responding to any negative reviews.

Now, Janice Dora, the operation's newly assigned general manager, was asking Roz to also respond to some guests who had posted positive reviews.

"I know it's going to take some time," said Janice, "but I think it's time well spent. If we only respond to our negative reviewers, then we're basically not valuing our happiest customers, only the angriest ones. I think our review scores are increasingly important. If it turns out that responding to some of the positive reviews takes you away from your other duties, then I think it's worth it to get you some help."

(Continued)

> Assume you were Janice. Do you think the additional labor costs that may be incurred by providing Roz with additional help should be considered a marketing cost or labor cost? How convinced would you be that the expense involved in responding to positive reviews is a legitimate marketing cost? Explain your answer.

Key Terms

Paid social (advertising)	Fake review	Arithmetic average
Call to action	Buyer's remorse	

Operator's 10-Point Tactics for Success Checklist

Evaluate your need for, and the current status of, each of the following operational tactics. For those tactics you think are important, but not yet in place, develop an action plan for its implementation including who will be responsible for the tactic's completion and the target date by which it should be completed.

Tactic	Don't Agree (Not Done)	Agree (Done)	Agree (Not Done)	If Not Done	
				Who Is Responsible?	Target Completion Date
1) Operator understands the reasons for the growing popularity and importance of UGC sites.	____	____	____		
2) Operator has a system in place for monitoring increases and decreases in the popularity of various UGC sites.	____	____	____		
3) Operator has assessed the wisdom of including paid social advertising as a major component of their marketing plan.	____	____	____		
4) Operator has reviewed the key actions to be employed as they plan and implement their paid social efforts.	____	____	____		

Tactic	Don't Agree (Not Done)	Agree (Done)	Agree (Not Done)	If Not Done	
				Who Is Responsible?	Target Completion Date
5) Operator recognizes the importance to their online reputation of the content and ratings posted on popular UGC sites featuring guest reviews.	___	___	___		
6) Operator has identified UGC sites featuring user reviews that are most commonly viewed by members of their target markets.	___	___	___		
7) Operator recognizes the steps to take to increase the number of reviews of their operation that are posted on UGC sites featuring user reviews.	___	___	___		
8) Operator recognizes the steps to take to increase the scores that their operation receives on UGC sites featuring user reviews.	___	___	___		
9) Operator understands the importance of promptly responding to negative reviews posted on popular UGC sites featuring user reviews.	___	___	___		
10) Operator understands the importance of professionally written responses to negative reviews posted on popular UGC sites featuring user reviews.	___	___	___		

12

Assessment of Marketing Efforts

What You Will Learn

1) The Importance of Marketing Effort Evaluation
2) How to Use Marketing Evaluation Tools
3) Considerations for Preparing Next-Period Marketing Plans

Operator's Brief

In this chapter, you will learn that foodservice operators must regularly evaluate their marketing efforts, and this is usually done when the time addressed in the operation's marketing plan has concluded.

When assessing the effectiveness of marketing efforts, it is important to examine the strategies indicated in previous marketing plans to assess how well previously identified goals were met. In addition, the marketing tactics used to support selected strategies are evaluated to determine if the tactics were implemented and, if so, the extent to which they were effective. These marketing tactics can be assessed by determining whether a tactic was implemented, examining the cost of its implementation, and assessing its effectiveness.

As the impacts of marketing efforts are assessed, foodservice operators analyze several financial and nonfinancial metrics (measuring standards). Important financial metrics include those related to total revenue generation, customer counts, guest check averages, and revenue generation by service style.

Important nonfinancial metric assessments are also increasingly important to the success of a foodservice operation and should also be reviewed. These include measurement of website traffic, evaluation of social media efforts, and an examination of scores on user-generated content (UGC) sites.

A thorough evaluation of an operation's marketing efforts in one time period helps operators as they plan their marketing efforts for the next time period. One important evaluation tool used in this process is a SWOT analysis.

A SWOT analysis is a formal evaluation of an operation's strengths, weaknesses, opportunities, and threats. Information learned from a SWOT analysis provides foodservice operators with information that will help them create next-period marketing plans and to consider their future marketing and profitability goals.

CHAPTER OUTLINE

The Importance of Evaluating Marketing Efforts
    Assessment of Marketing Strategies
    Assessment of Marketing Tactics
Marketing Evaluation Tools
    Tools for Financial Evaluation
    Tools for Nonfinancial Evaluation
Preparing Next-Period Marketing Plans
    SWOT Analysis
    Setting Next-Period Strategies and Goals
Summary

The Importance of Evaluating Marketing Efforts

The efforts used to market a foodservice operation are many and varied. As explained in Chapter 5, these efforts are formalized in an operation's marketing plan that indicates the marketing efforts to be undertaken, when they will be undertaken, and who will complete them.

When the time covered by the marketing plan has passed, effective foodservice operators evaluate the results of their marketing efforts. This involves assessing the extent to which strategies (goals) identified in the marketing plan were addressed and an evaluation of the tactics used to attain them.

Several factors can impact the results obtained by an operation's marketing efforts. For example, an operation may have a very effective marketing program. However, if it fails to deliver on its promises when customers visit the operation, long-term financial results may not be achieved. Alternatively, an operation may do only a limited amount of marketing, and its customer counts might continually increase because the product quality and service levels of the operation are high.

The reasons for undertaking a detailed examination of an operation's marketing results are many and include:

✓ Gaining a better understanding of a target market's behavior
✓ Allowing for a quantitative assessment of key **revenue metrics** that are impacted by marketing
✓ Providing a systematic assessment of online reputation and brand awareness
✓ Making better decisions for future marketing efforts

Key Term

Revenue metric: A standard for measuring or evaluating data based on its dollar amount or change in dollar amount. Also known as a "financial metric."

Internal and external factors can impact an operation's ability to achieve marketing plan objectives. A regularly planned and objective assessment of marketing efforts is required so operators can identify marketing activities that worked well and others that did not work as well. This information is essential if managers are to improve their marketing plans from one time period to the next.

As shown in Figure 12.1, four important cyclical steps are needed to plan and implement a revised marketing plan. The operator must (first step) assess the current marketing plan results and, (second step), modify marketing strategies and

Figure 12.1 Marketing Plan Assessment and Improvement

tactics as needed. It is then possible to prepare (Step 3) and implement (Step 4) the next period's marketing plan. At some point that marketing plan will again be revised, and the cycle in Figure 12.1 will be repeated.

When assessing the marketing plan's results and beginning to plan future marketing plans as needed, operators will benefit from studying the results of the marketing strategies and marketing tactics detailed in the then-current plan.

Assessment of Marketing Strategies

Chapter 5 defined a marketing strategy (goal) as an objective a foodservice operation wishes to achieve, and the best of these objectives are both realistic and measurable. Likewise, when they assess the achievement of market strategies, operators compare planned results with actual results.

For example, assume an operator identified an objective to increase takeout business by 10% during the time span of the marketing plan. That objective can be quantifiably measured and, when the time addressed in the marketing plan is concluded, the degree to which the objective was achieved can be assessed. If the objective was achieved, the operator can look to the specific tactics that were successfully used to support the objective. If the objective was not achieved, the operator can then assess the individual tactics put in place to help achieve the objective and judge which tactics were successful and identify others that were not, and as a result, need replacement or revision.

Sometimes a marketing objective will not be met despite an operator's best marketing efforts to do so. One reason can be that external factors impact the operation's success.

This was clearly illustrated by the COVID-19 pandemic that began in early 2020. Foodservice operators developing their marketing plans for that year had no idea the pandemic would result in foodservice sales being down for the year throughout the industry. Note: The National Restaurant Association reported that restaurant and foodservice sales in 2020 were $240 billion (27%) below its 2020 pre-pandemic forecasts. Also, more than 110,000 eating and drinking establishments closed that year, either temporarily or permanently.[1]

There are numerous external factors that can contribute to a foodservice operation not achieving marketing goals. A depressed economy, layoffs by a major employer in an operation's marketing area, and nearby road construction are examples. Others, including the opening of a competitive operation and severe weather such as hurricanes or floods, can also impact an operation's revenue-generating abilities and ability to meet marketing goals.

---

1 https://www.cnbc.com/2021/01/26/restaurant-industry.html retrieved June 30, 2022.

As they assess marketing success, operators must keep external factors in mind and make a reasonable determination about reasons for unattained objectives.

Assessment of Marketing Tactics

Marketing tactics are the specific steps and actions operators undertake to achieve their marketing objectives. An assessment of marketing tactics is useful regardless of whether the objectives to be attained were achieved or not achieved.

When assessing specific marketing tactics, operators should consider the answers to three important questions:

1) Was the tactic implemented?

Foodservice operators identify tactics to be implemented in support of their goals as they create marketing plans. In some cases, a tactic is not implemented or is only partially implemented during the time period covered by the marketing plan.

To illustrate, assume that a foodservice operator identified two sales calls per month to be undertaken as an important tactic when the annual marketing plan is developed. At the conclusion of the marketing plan's established time period, it is determined that the operator made only one personal sales call per month. In this case, partial (not full) implementation of the tactic may have impacted its effectiveness.

In some cases, a marketing tactic is implemented in a different time frame than originally designated. For example, assume a foodservice operator identified the implementation of up-selling training as an important tactic to attain a 5% increase in guest check average during the time period addressed by the marketing plan.

The marketing plan called for the up-selling training (see Chapter 8) to occur during the first quarter of the year. In fact, the training was not implemented until the third quarter of the year. In this example, while the tactic was implemented, the timing of the training did not optimize its effectiveness in helping the operation achieve a 5% increase in guest check average.

Identifying whether each tactic listed in the marketing plan was implemented is also a critical step in determining the tactic's effectiveness. It is not normally appropriate to, for example, examine a tactic that increased revenue by 7% more than planned and fail to examine another unimplemented tactic.

2) Was the tactic's actual cost consistent with its marketing plan cost estimate?

As marketing plans and budgets are developed, foodservice operators make estimates about the cost of implementing specific tactics included in their plans. In some cases, the actual cost of implementing the tactic is consistent with the original budget forecasts. In other cases, however, the cost of implementing the tactic may be more (or less than) originally forecast.

In many cases, foodservice operators undertake specific marketing tactics to increase guest counts, increase total revenue, and/or increase guest check average. Assessing the actual cost of implementing a tactic is important for determining the tactic's **marketing return on investment (ROI)**.

Key Term
_____

Marketing return on investment (ROI): The amount of benefit gained from a marketing activity compared to the initial cost of the activity.

A marketing activity's ROI is the difference between how much is spent on a specific marketing activity and how much revenue it generates. The formula foodservice operators use to calculate their marketing ROI is:

$$\frac{\text{Marketing revenue} - \text{marketing cost}}{\text{Marketing cost}} \times 100 = \text{Marketing ROI}$$

To illustrate, assume an operation implements a marketing tactic that cost the operation $28,000. The impact of the tactic was an increase in revenue of $98,000 during the time period covered by the marketing plan. In this example, the operator's ROI for this marketing tactic would be calculated as:

$98,000$ revenue $- \$28,000$ marketing cost $= \$70,000$
$\$70,000 \,/\, \$28,000 = 2.5$
$2.5 \times 100 = 250\%$

In this example, the operation achieved a 250% ROI on its marketing activity. Stated another way, for every dollar the operation spent implementing this marketing tactic, $2.50 in additional revenue was generated.

The impact resulting from the implementation of some marketing tactics cannot easily be measured in dollars or ROI. For example, if a foodservice operator elects to hold a fundraiser for a local charity and, in doing so, expends $1,000 in product costs for attendees of the event, the resulting impact on the operation's reputation and the goodwill generated by hosting the event cannot easily be measured.

The measurement of ROI allows operators to compare the revenue impact of various marketing tactics. For example, assume that an operator finds that, when an investment is made in online marketing efforts, the return on investment is 300%. When the organization invests in local newspaper ads, the resulting ROI is 125%. In this example, it may make good sense for the operator to increase the amount of online advertising and reduce the amount of paid newspaper advertising because of the resulting ROIs.

Regardless of whether a marketing activity's impact can or cannot be readily measured, foodservice operators should establish the cost they incurred to implement their marketing tactics. Doing so allows them to make better decisions going forward about whether such tactics should be implemented again in the future, and it also provides data for improving cost estimates in future marketing plans.

3) Was the tactic effective?

In many respects, this is the most important question of all, and it can also be one of the most challenging questions for an operator to answer. In some cases, operators can identify quantitative measures of advertising effectiveness. For example, they can monitor total revenue generation, count the number of new customers signing up for guest loyalty programs, measure total website traffic, and/or assess website click-through rates (CTRs) (see Chapter 9).

While each operation is different, in many cases advertising effectiveness can be assessed by:

Monitoring guest counts

One relatively easy way to determine if advertising is working is to record changes in guest traffic by counting the number of guests who are served or who visit a website. When operators monitor traffic before they start an ad campaign, they will then have a basis for comparison. It is also helpful to ask new customers how they heard about the business.

Coding coupons

When possible, producing ads with a coupon that customers can redeem for a purchase discount is a good way to measure impact. When the coupons are coded with a unique identifier an operator can determine which ad, website, or publication generated the best results.

Offering incentives

Foodservice operators can offer incentives for guests in their advertising efforts. For example, an operation may offer incentives for customers to communicate that they are responding to an ad by stating "Mention this ad and get a 10% discount on your meal purchase." When incentives are advertised on a website, in local publications, or on radio operators will know the impact of the advertising medium based on the number of customer responses to the incentive.

Monitoring sales

Monitoring sales means making careful assessments before, during, and after an advertising tactic is undertaken. This is an important activity, and the ways to do it are addressed in detail in the next section of this chapter. What foodservice operators must recognize, however, is that advertising may have a cumulative or delayed effect, and advertisement-driven sales may not be identified immediately. This is especially so when an advertising tactic is designed to support a goal of increasing brand awareness or enhancing the reputation of an operation in its local community.

What Would You Do? 12.1

"Well, I think the promotion was a great success," said John, the co-owner of Guac and Roll, a food truck featuring California-style Mexican menu items.

John was discussing a recent advertising promotion that he and his partner Armando had implemented to increase awareness of where their food truck would be parked on the Fourth of July weekend.

"Last year on the 4th we did $12,000 for the entire weekend," continued John, "with this year's promotion our sales this year were $15,000. That's a $3,000 increase in revenue, and the promotion only cost us $1,000 in advertising expenses, so we have a 300% return on our investment!"

"I agree," said Armando, "but we don't have a 33% profit margin. The way I see it, after we paid for the ad, we brought in an additional $2,000 in sales, but our food, labor, and other associated costs were about $2500. That seems like a loss of $500 to me!"

Assume you were advising John and Armando. Do you think a 300% ROI for this advertising program means it was a success? Would you advise them to run a similar advertising campaign next year? Explain your answer.

Marketing Evaluation Tools

Foodservice operators have several tools available as they evaluate their marketing efforts. Some of these tools relate to financial evaluations, while others are **nonfinancial metrics**.

Whether financial or nonfinancial, the use of well-defined metrics provides foodservice operators with an objective measure of their overall marketing performance. The metrics used to evaluate a foodservice operation's marketing efforts should be easy to understand and calculate.

Key Term

Nonfinancial metric: A standard for measuring or evaluating something, especially one that uses nondollar figures or statistics. Also commonly referred to as a "key performance index (KPI)."

One good way for foodservice operators to consider the metrics they use to assess their overall marketing efforts is to view them as either being a tool for financial evaluation or for nonfinancial evaluation.

Tools for Financial Evaluations

Chapter 5 defined "revenue variance" as the difference (in dollars) between what a foodservice operator estimated would be generated from a specific marketing strategy and what the strategy actually generated. When foodservice operators evaluate the impact of their overall marketing efforts, they can use the same revenue variance formula:

$$\text{Actual revenue} - \text{Estimated revenue} = \text{Revenue variance}$$

Since revenue variance (in dollars) is partially a function of operational size, many operators prefer to utilize a percentage variance to assess their overall marketing efforts. Recall from Chapter 5 that the formula used to calculate a revenue percentage variance is:

$$\frac{\text{Actual revenue} - \text{Estimated revenue}}{\text{Estimated revenue}} = \text{Percentage variance}$$

To illustrate, assume a foodservice operation implemented a marketing plan for a 12-month period. The marketing plan's revenue forecast indicated that the operation would move from $2,000,000 in sales in the previous year to $2,200,000 in the year addressed by the marketing plan. This is an increase in revenue of 10% ($2,200,000 − $2,000,000)/$2,000,000 = 10%).

At the end of the 12-month period, it was determined that the operation achieved actual sales of $2,450,000. In this example, the operation's revenue variance would be calculated as:

$$\$2,450,00 \text{ Actual revenue} - \$2,200,000 \text{ Estimated revenue}$$
$$= \$250,000 \text{ Revenue variance}$$

In this example, the operation's actual percentage variance is calculated as:

$$\frac{\$2,450,000 \text{ Actual revenue} - \$2,200,000 \text{ Estimated revenue}}{\$2,200,000 \text{ Estimated revenue}} = 11.4\% \text{ (rounded)}$$

In this example, the operation exceeded its estimated revenue increase by 1.4% (11.4% actual change − 10.0% estimated change = 1.4%).

Understanding how to use these two basic variance equations provides foodservice operators with tools needed to undertake several additional and important performance assessments including:

✓ Evaluation of Changes in Revenue Generation Totals
✓ Evaluation of Guest Counts
✓ Evaluation of Guest Check Averages
✓ Evaluation of Changes in Revenue Generation by Service Method

Evaluation of Changes in Revenue Generation Totals

In nearly all cases, the intent of marketing efforts is to increase total revenue. Changes in total revenue generation should be a major factor when assessing the quality of a foodservice operation's marketing efforts. A close look at the total revenue formula:

$$\text{Guest count} \times \text{Guest check average} = \text{Total revenue}$$

reveals there are several possible ways an operation might generate an increase (or decrease) in total revenue:

1) More guests were served at the same check on average
2) The same number of guests was served but at a higher guest check average
3) More guests were served, and at a higher guest check average
4) Fewer guests were served, but at a higher guest check average

As a result, the proper way to evaluate changes in total revenue is to evaluate both guest counts and guest check averages.

Evaluation of Guest Counts

Effective marketing should increase an operation's guest count. While many financial metrics are important, customer count (a nonfinancial metric), is likely the single most important metric to be evaluated by an operation. To maintain its current revenue stream, and increase it, in most cases, the operation must serve additional guests each year. As a result, an assessment of an operation's marketing efforts begins with an assessment of guest counts.

Modern point of sale (POS) systems are programmed to report the amount of revenue generated in a selected time period, the number of guests served in that period, and the average sale per guest (guest check average). Therefore, the POS system provides the guest count data operators require to assess changes in guest counts, and this assessment must be carefully undertaken.

To illustrate why this is important, consider Stephan Bart, an operator who launches an aggressive advertising campaign in the first quarter of his marketing year. At year's end, he determined that his operation served 795,000 guests, up from 750,000 guests the previous year.

Using the formulas below, this operation's change in guest count is calculated as:

$$\text{This year's guest count} - \text{Last year's guest count} = \text{Guest count change}$$

In this example that would be:

$$795,000 \text{ guests} - 750,000 \text{ guests} = 45,000 \text{ guests}$$

Using the formula here, the operation's percentage change in the guest count would be calculated as:

$$\frac{\text{Guest count change}}{\text{Last year guest count}} = \% \text{ Change in guest count}$$

In this example, that would be:

$$\frac{45,000 \text{ guests}}{750,000 \text{ guests}} = 0.06 \text{ or } 6\%$$

It appears Stephan's operation realized successful marketing efforts because it increased its guest count by 6% on a year-over-year (YOY) comparison basis.

However, consider the guest count data shown in Figure 12.2, which presents the operation's actual guest counts on a quarterly (not annual) basis.

The data shows Stephan's marketing campaign was effective in the first quarter: his marketing tactics drew additional customers to the operation. However, a closer look at the data also shows that, in the third and fourth quarters, customer counts were lower than in the prior year. This closer analysis of his guest count data should concern Stephan because there is a drop off in customer counts at an increasing rate.

Time Period	This Year	Last Year	Difference
First quarter	225,000	180,000	45,000
Second quarter	205,000	175,000	30,000
Third quarter	190,000	200,000	−10,000
Fourth quarter	175,000	195,000	−20,000
Annual total	**795,000**	**750,000**	45,000

Figure 12.2 Stephan's Actual Guest Counts

This example also illustrates the dangers inherent in collapsing data into segments that are too large. Stephan's annual data does not provide as useful a picture of what is happening as would an analysis of his quarterly guest data.

Evaluation of Guest Check Averages

Recall that the formula used to calculate a foodservice operation's total revenue is:

$$\text{Guest count} \times \text{Guest check average} = \text{Total revenue}$$

This formula indicates that an operation's guest check average is a critical part of its total revenue generation. Guest check average in a foodservice operation is primarily impacted by its menu prices and the up-selling efforts of its service staff.

Foodservice operators normally want to see an increase in guest check average from one marketing period to the next. The reason: This positive change indicates guests like the products they are offered, and it is likely that service staff are effective with up-selling efforts.

Two methods are available to properly evaluate guest check average changes. The first method assumes there has been no variation in menu prices from the prior marketing period. The second method considers the impact of menu price variations when calculating guest check average changes.

Guest Check Average Change with No Variation in Menu Prices

When there have been no changes to menu prices from one marketing period to the next, the calculation of guest check average change is relatively simple and straightforward. To illustrate, consider the case of Jhanell Rakunia, who operates a gourmet deli/sandwich shop featuring extremely high-quality and unique ingredients.

One of Janelle's goals for the just-completed marketing period was to increase her operation's guest check average by 7%. Her actual guest check average was $18.78, which is an increase from the $16.95 in the prior marketing period. Using the formula below her operation's change in guest check average would be calculated as:

$$\text{This period guest check average} - \text{Last period guest check average}$$
$$= \text{Guest check average change}$$

In this example that would be:

$$\$18.78 - \$16.95 = \$1.83 \text{ Guest check average change}$$

Using the formula below, Jhanell's percentage change in guest check average would be calculated as:

$$\frac{\text{Guest check average change}}{\text{Last period guest check average}} = \% \text{ Change in guest check average}$$

In this example that would be:

$$\frac{\$1.83}{\$16.95} = 0.108 \text{ or } 10.8\%$$

From these calculations, it appears Jhanell's marketing efforts were successful because she exceeded her marketing plan's target check average increase of 6.0% by 4.8% (10.8% actual increase − 6.0% targeted increase = 4.8%).

Guest Check Average Change with Variation in Menu Prices

In 1968, McDonald's introduced the Big Mac to its menu nationwide. In that year, the Big Mac sold for $0.45. Today's Big Mac sells for much more than $0.45. Therefore, attempting to make a direct comparison between the dollar revenue generated by Big Mac sales in 1968 and those generated today are not useful because the selling price of Big Macs is different. This example illustrates that menu prices must periodically be increased to address issues such as increased product, labor, and operating costs.

When operators want to assess changes in guest check averages between marketing periods during which menu prices were *not* the same, they must consider menu price variation in their calculations. To illustrate this factor, let's return to Jhanell's operating results above (and recall her guest check average was 10.8% higher in the current marketing period when compared to the previous period).

Assume now, however, that when beginning this marketing period Janelle had raised her prices by 5%, and she now wants to determine her actual guest check average increase (after adjusting for changes in menu prices).

This is a three-step process.

Step 1. Increase the prior-period (last-year) guest check average by the amount of the price increase.

Step 2. Subtract the result in Step 1 from this period's guest check average.

Step 3. Divide the difference in Step 2 by the value of Step 1.

In this example, Jhanell would follow the three steps to determine her "real" (after adjustment) change in guest check average. The procedure would be:

Step 1. $16.95 (prior period guest check average) × 1.05 = $17.80

Step 2. $18.78 (this period's guest check average) − $17.80 = $0.98

Step 3. $0.98 ÷ $17.80 = 0.055 or 5.5%

Figure 12.3 details the results Jhanell achieved when her 5% menu price adjustment is included.

	Last Period	Adjusted Sales (Last Period × 1.05)	This Period ($)	Variance ($)	Variance (%)
Guest check average	$16.95	$17.80	18.78	0.98	+ 5.5

Figure 12.3 Jhanell's Guest Check Average Comparison with 5% Menu Price Increase

Note that, in the data reviewed in Figure 12.3, Jhanell's overall guest check average did increase by 10.8% during this marketing period. However, the actual "after-menu-price-adjustment" was 5.5%, and this is just short of her increased goal of 6.0%. Figure 12.3 indicates it is important to consider making menu price adjustment calculations when evaluating revenue generated in one marketing period versus another period.

Evaluation of Changes in Revenue Generation by Service Method

A foodservice operation's guests can frequently choose between different service or delivery styles when making their purchases. When available, it is useful for operators to evaluate changes in revenue based on the service styles they offer to guests.

To illustrate why this is important, consider a full-service restaurant and lounge that markets its takeout and delivery service and the dine-in service that is also available. As shown in Figure 12.4, at the conclusion of this marketing period, the operator's revenue was up 1.6% from the previous marketing period.

A close examination of the operator's revenue generation data reveals that sales generated from dine-in customers were 14% less than in the prior marketing period. Since fewer guests chose the in-dining option, lounge beverage sales were also down a similar amount. Takeout and delivery service in this operation, however, showed significant increases of 29.2% and 29.8% respectively. In this

Service Style	Revenue This Period ($)	Revenue Last Period ($)	Variance ($)	Variance (%)
Dine in	103,500	120,300	16,800	−14.0
Lounge	32,100	37,500	5,400	−14.4
Take out	66,525	51,500	15,025	29.2
Delivery	47,900	36,900	11,000	29.8
Total	250,025	246,200	3,825	1.6

Figure 12.4 Full-Service Restaurant with Lounge Revenue Change

example, it is easy to see that an assessment of revenue generation by service method will reveal more information to this operation's managers than would an evaluation of the operation's overall revenue generation.

The proper evaluation of changes in revenue generation may also require operators to consider special factors. For example, a fast-casual foodservice operation might be located across the street from a professional basketball arena. If the operation compared sales from this April to last April, the number of April home games played by the basketball team might differ. Then a valid conclusion about guest count increases or decreases could only be made if the number of home games is known because that number would likely impact April revenue levels.

Similarly, if a foodservice operation is open only Monday through Friday, the number of these operating days contained in two given marketing periods may be different. Then percentage increases or decreases in sales volume must be based on average daily sales, rather than on total sales.

To illustrate, consider a food cart operation that serves Chicago-style hotdogs. The cart operates in the city center of a large metropolitan area and is only open Monday through Friday. In October of this year, the stand was open for 21 operating days. Last year, however, because of the number of weekdays and weekend days in October, the stand operated for 22 days.

Figure 12.5 details the comparison of sales for the food cart assuming no increase in menu selling prices this year when compared with last year's selling prices.

Upon first examination October sales this marketing period are 1.2% *lower* than last year. However, average daily sales are up 3.6%.

Were October sales up or down? Clearly, the answer must be qualified in terms of monthly or daily sales. For this reason, effective foodservice operators must consider all relevant factors impacting revenue before making determinations about an operation's overall sales direction. Note: This assessment also illustrates a point made earlier about collapsing data too extensively. In this example, the food cart's monthly data is not as accurate as its daily data.

For many operators, special revenue-impacting factors can include the number of operating meal periods or days in the marketing period. Other factors include

	Last Period	This Period	Variance	Variance (%)
Total sales (October)	$17,710	$17,506	$−204	−1.2
Number of operating days	22 days	21 days	1 day	
Average daily sales	$805.00	$833.62	$28.62	+3.6

Figure 12.5 Food Cart Sales Data

menu price changes, total guest counts, guest check averages, and holidays and special events during the period. Only after carefully reviewing all relevant information can operators truly assess the degree to which their operation's sales are increasing or decreasing due to their marketing efforts.

Find Out More

In many cases, those who start their own commercial foodservice operations consider themselves to be either "Food" people or "People" people.

Those who consider themselves to be "Foodies" (food people) are extremely interested in the products they serve and the way these foods are prepared. "People" persons are typically drawn to the hospitality industry because of their love of service to others.

Whether they are "food" people, or "people" people, all foodservice operators must have a solid understanding of the financial management of their businesses. While foodservice operators need not be accountants to succeed, they do need to understand how their financial operating data is compiled and presented. One good source of this information is the Wiley-published book *Managerial Accounting for the Hospitality Industry*, written by Dr. Lea Dopson and Dr. David Hayes.

To review this and other currently available software programs and written publications that assist operators in understanding the financial management and reporting aspects of their businesses, enter "financial accounting for restaurants" in your favorite browser and view the results.

Tools for Nonfinancial Evaluation

In addition to financial metrics developed at the end of a marketing period, operators must also measure important nonfinancial metrics. For example, digital marketing metrics (see Chapter 4). measure the success of an operation's online marketing efforts and the quality of its online reputation.

While the specific digital marketing metrics of importance vary by operation, all operators should regularly undertake three important assessment tasks at the end of a marketing period:

✓ Measurement of Website Traffic
✓ Evaluation of Social Media Efforts
✓ Evaluation of Scores on User-generated Content (UGC) Sites

Measurement of Website Traffic

The host of a foodservice operation's website can provide much information about activity on the site. The specific data analyzed by operators varies based on a website's features (see Chapter 9), but important digital marketing metrics for websites that should be regularly assessed by operators include:

Number of visitors

The number of visitors to an operator's website is a key nonfinancial metric because attracting viewers to a website is often the first step in developing a relationship with them. Operators should monitor the number of visitors to their sites for spikes (fast increases) or sudden decreases in traffic. Increases in traffic likely come from website improvements and/or external marketing efforts. Decreases in traffic may indicate a website that is ineffective in drawing viewers.

In addition to the total number of visitors to a site, operators should also monitor the source of visitors who may reach a site with, for example, an organic search on Google, the operation's social media channels, or from a third-party referral link.

Bounce rate

Chapter 9 defined "bounce rate" as the percentage of visitors who view only one page on a website before leaving it. These short-term visitors may leave a website by clicking an external link, pressing their browser back button, typing another URL into their browser, or simply closing their browser window or tab.

In most cases, a poor bounce rate means a poor user experience and ultimately lower sales from users who leave a website before learning about an operation and its product/service offerings. Therefore, monitoring the bounce rate can improve a website's performance.

Average time on page

Generally, the longer website visitors stay on the site, the more engaged they will be. Monitoring the average time on a page lets operators know if their website visitors are finding information of interest that prompts them to read or view the site.

For sites with video, monitoring the average time on the page enable operators to know more about their visitors' viewing habits. For example, if a posted video is eight minutes long, but the average visitor stays on a site for only two to three minutes, most of the video is not being viewed, and the operator may want to shorten the video's length.

Click-through rate

A website's CTR (see Chapter 9) measures how successful a specific ad or call to action has been in capturing a website viewer's attention. In general, the higher the CTR the more successful the ad or call to action has been in communicating a specific marketing message to a website visitor.

Evaluation of Social Media Efforts

An operation's social media efforts can be considered successful if there is an increasing number of shares, likes, subscribers, retweets, or comments about the information posted to the operation's social media sites. However, a large number of followers who do not become customers is less desirable than a smaller audience that engages in sharing, commenting on, and liking the information posted on an operator's social media sites (see Chapter 11).

Operators should recognize that the popularity of any specific social media site may increase or decrease. Therefore, operators must regularly monitor engagement levels on each of their social media sites.

Technology at Work

Few foodservice operators have the skill, knowledge, and time required to properly monitor and interpret the results of all their social media marketing efforts.

Effective social media monitoring means tracking hashtags, and keywords and mentions about an operation. By monitoring this type of data, foodservice operators gain both quantitative information (metrics) and qualitative information that may provide inspiration for future posts and marketing strategies.

To view some of the organizations formed to assist foodservice operators in the monitoring of their social media efforts, enter "social media monitoring for restaurants" in your favorite search engine and view the results.

Evaluation of Scores on User-Generated Content Sites

Chapter 11 detailed the importance of a foodservice operation's rating scores on user-generated content sites. At the conclusion of each marketing effort, operators should assess their rating scores on the major review sites. A properly executed marketing plan should have the effect of improving, or at least maintaining, an operation's numerical review scores.

User reviews are especially important for small operations seeking to boost their online reputation and grow their customer base. Increasing numbers of customers (and especially Millennials and those in Generation Z) look to the presence of a foodservice operation's online brand when deciding whether their personal needs in terms of quality, dining experience, menu, and price will be met.

Monitoring evaluation scores on-user generated content sites are a critical part of the marketing effort because negative online reviews can have a devastating effect on an operation's reputation. Studies show that 3.3 is the minimum star rating for business consumers would engage with, and only 13% of consumers consider using a business that has only a one- or two-star rating.[2]

In addition, those same studies show that four out of five consumers have changed their minds about visiting a recommended foodservice operation after reading negative online reviews. Note: To learn about specific strategies and tactics operators can use to improve their scores on user-generated content sites, see Chapter 11.

Preparing Next-Period Marketing Plans

Recall from Chapter 1 that the overall goal of a marketing effort is to attract customers from a target market, satisfy these guests' wants and needs, and thus encourage an ever-expanding customer base. Achieving that goal with the help of an effective marketing plan requires operators to assess the effectiveness of their plans, carefully identifying what worked and what did not. When they do so, they will better know about the effectiveness of their past efforts and where best to devote their efforts and marketing funds in the future.

The evaluation of a specific marketing plan should begin as soon as the deadline date for that plan has passed. In addition to assessing the strategies and tactics that were implemented, foodservice operators must carefully assess any new internal or external factors that may affect their future marketing efforts. One good way to do that is through the implementation of a SWOT analysis.

SWOT Analysis

A **SWOT analysis** is a popular method for examining an operation's position in its marketplace to improve its overall marketing efforts. As shown in Figure 12.6, SWOT is short for "Strengths, Weaknesses, Opportunities, and Threats."

SWOT analysis asks operators to examine their businesses and carefully consider:

✓ What are our **Strengths**?
✓ What are our **Weaknesses**?
✓ What are our **Opportunities**?
✓ What are the **Threats** we face in the future?

Key Term

SWOT analysis: Strengths, Weaknesses, Opportunities, and Threats analysis; a systematic approach to assessing an operation's marketing strategies and tactics planning.

---

2 https://www.qualtrics.com/blog/online-review-stats/ retrieved July 30, 2022.

Figure 12.6 SWOT Analysis

A SWOT analysis identifies *strengths* and considers how to increase and more fully utilize them. It identifies *weaknesses* and considers how to overcome them. Existing *opportunities* are explored to take advantage of them more fully. Finally, *threats* to an operation's long-term success are identified so they can be better addressed and attained.

To assess an operation's strengths, operators review data to identify extremely popular menu items, unique service offerings, and/or the operation's location. Weaknesses of an operation may be identified by reviewing negative comments posted on UGC review sites and by seeking input from service staff about verbalized customers' likes and dislikes.

Opportunities identified in a SWOT analysis may include changes that will allow an operation to cost-effectively gain a competitive edge, identify new target markets, or expand current capacity. In most cases, opportunities are external factors in the business environment that are likely to contribute to greater success.

To cite some examples, operators should consider if there are dining trends emerging that will encourage guests to buy more of what the operation is selling. Are there upcoming local events that the operation may be able to take advantage of to grow sales? Are there upcoming changes to the local competitive or economic environment that might impact the operation? In some cases, very specific opportunities may be identified. For example, if an economic trend in an area is toward two-income families with small children, such families typically have less time to cook, and thus special promotions may be targeted directly at this increasingly large demographic.

The identification of threats in a SWOT analysis can include the opening of competitive operations, rising operating costs, or difficulty attracting and retaining staff. Operators assessing potential threats may also want to consider whether their suppliers will always be able to provide needed raw ingredients at affordable prices. Could future developments in technology change how the operation must do business? Is consumer behavior changing in a way that could negatively impact the business? All of these examples and more could pose threats to the long-term profitability of an operation.

A SWOT analysis may reveal factors that cause an operation's revenue to only be equal to previous marketing periods or even to decline. Examples include the impact of weather, the opening of a new competitor's operation, and changes in consumer preferences. Other examples include poor product quality or service delivery, and the potential need for the business to upgrade its physical facility or technological capabilities.

In some cases, a SWOT analysis may reveal that an economic condition threatens the revenue stream of a business. In these cases, operators must recognize that economic downturns can be either short- or long-term. A revenue decline based on short-term economic conditions might simply be ignored because the duration of the condition could make overreaction a bigger threat than the economic condition.

Consider, for example, a foodservice operation located near a manufacturing facility that was built five miles from the closest city. In this operation, many breakfast and lunch customers work at the facility. Slowing sales of this manufacturer's products have resulted in an unplanned plant shutdown and employee layoff that will last two weeks.

Despite the inevitable and significant impact on volume, the best advice for this operation's marketing team is likely to stay the course. In this case, the cost of publicizing significant changes to the operation's pricing or marketing structure likely outweighs any incremental revenues these actions might generate.

Similarly, assume an operation located on an interstate highway exit learns that highway repairs need to be done that will close the exit leading to it for 60 days. In this scenario, it is unlikely that any short-term changes in strategic marketing will make a significant difference in the operation's next 60 days' business volume. Rather, some staffing changes (reductions) in the operation may be needed for the next two months.

Of course, factors that make a significant short-term impact on a local or regional economy can linger. In some cases, it may not be completely clear when or whether the impact of economic difficulty will cease, moderate, or get worse.

Figure 12.7 lists some common shorter- and longer-term economic factors that can significantly affect an operation's ability to maintain its historic revenue levels.

Shorter-Term Threats	Longer-Term Threats
Increased costs of travel/fuel	Terrorism attack or threat
Sensationalized local event (major crime, oil spill, hurricane, and the like)	Prolonged economic recession in the local or regional area
Temporary plant or office closings	Permanent major plant or office closings
Labor unrest/unionization efforts	Changing consumers' demographics, trends, and preferences
Short-term health epidemic or pandemic (e.g. Norovirus outbreak and H1N1 flu, ecological disasters such as oil spills on beaches and hurricanes)	Longer-term health epidemic or pandemic (e.g., COVID-19 virus)
Product and/or labor shortages	Persistently high levels of inflation that reduce guests' purchasing power

Figure 12.7 Potential Economic Threats to Foodservice Businesses

It is important to recognize that a sales volume problem that was *not* caused by poor marketing or product/service quality can rarely be solved by additional marketing. The best marketing advice for operators facing short-term economic threats to their businesses may simply be to focus on product and service quality and avoid any radical change in marketing efforts.

Find Out More

Foodservice operators increasingly identify an inability to attract and retain hourly employees as they undertake SWOT analyses.

Additionally, when assessing their staff, some operators focus their attention mainly on their salaried supervisors and managers. However, successful operators know that long-term success will be due in large part to the efforts of their hourly employees, and it is among these workers that the most serious staff shortages are encountered.

Operators who create welcoming and positive work environments for their hourly-paid employees will likely notice improved employee performance, higher job satisfaction, lower turnover rates, and the ability to attract talented hourly workers in the future.

Regardless of the number of hourly workers employed by a foodservice operation, all such workers should be offered:

Fair compensation
Flexible schedules
Competitive benefits
Opportunities for advancement
The ability to be heard when they voice job-related concerns

If employees feel they cannot talk honestly to their supervisors or manager if their jobs become overwhelming to them, they are likely to seek employment elsewhere. Also, when hourly employees know their managers truly care about them, they are less likely to leave.

Maintaining positive relationships with hourly staff cannot be an occasional activity. Fostering good relationships with hourly employees should be part of every operator's daily priorities. To review suggestions on attracting and retaining a high-quality hourly workforce, enter "how to attract and retain hourly workers" in your favorite browser and view the results.

Technology at Work

To properly prepare a SWOT analysis foodservice operators must be realistic about assessing themselves and their competitors. When identifying their strengths, operators should only indicate those areas in which they truly excel. Weaknesses must be identified so they can be eliminated or turned into strengths.

Opportunities are those areas in which an operation can use their advertising mix to increase guest counts, increase guest check averages, or improve one or more digital marketing metrics. Threat identification is important because such threats can often be addressed and managed before they can do significant damage to an operation's revenue stream or reputation.

The actual preparation of a formal SWOT analysis is made easy today because of inexpensive software programs created for this purpose. To view the features and costs of these programs enter "create a SWOT analysis" in your favorite search engine and view the results.

Setting Next-Period Strategies and Goals

After completing a review of their previous marketing plan's results (assessments of the past) and conducting a SWOT analysis (assessments of the future), those responsible for creating an operation's marketing plan are prepared to determine their next marketing period strategies and goals.

Recall from Chapter 5 (see Figure 5.2) that Step 1 in Market Plan Development was presented as "Determine Market Goals." The setting of "Next Marketing Plan Period" strategies and goals follow the same steps; beginning with "Determine Marketing Goals."

As shown in Figure 12.8 all five steps in the development of the initial market plan will apply to setting next-period strategies and goals, as well as the tactics to be used to achieve them.

In addition to the results achieved in the prior marketing period, and a SWOT analysis, the preparation of next-period marketing plans should always include an operator's careful assessment of **market trends** and **market potential**.

An assessment of market trends helps to ensure that an operator's view of target markets will change if the behavior of the target markets changes. A realistic assessment of market potential gives operators insight into new markets to explore and the probable future size of existing markets.

Key Term

Market trend: Quantifiable changes in the behavior of a target market such as what target members buy, how they buy, and when they buy.

Key Term

Market potential: An estimate of total dollars or other units of measure for a specific product including those sold by an operation's competitors.

Figure 12.8 Steps in Next-Period Marketing Plan Development

As they follow the steps in Figure 12.8, operators must always remember that their new marketing plan goals should:

1) *State an identified time frame*; for example: within 60 days, by October 1, or by the end of the marketing period.
2) *State a measurable result*; for example: by $3.50, by 15%, by one star, or by three ranking levels on a Google search result.

When they do so they will be establishing financial and nonfinancial metrics that can be used as they assess the effectiveness of their next-period marketing plan when it comes to its conclusion.

Summary

All foodservice operators must effectively market their businesses. This is true whether the operation is a full-service fine dining restaurant, a college foodservice, or a local food truck. To do so, they must focus on the needs and wants of customers, and they must tell target customers how their needs and wants will be satisfied by their operation.

As this book has detailed, properly marketing a foodservice business requires operators to:

✓ Understand the importance of marketing (Chapter 1)
✓ Identify target markets (Chapter 2)

✓ Create a unique marketing message (Chapter 3)
✓ Choose appropriate communication channels for marketing message delivery (Chapter 4)
✓ Prepare a marketing plan (Chapter 5)
✓ Recognize the importance of price in marketing (Chapter 6)
✓ Utilize the menu as a major marketing tool (Chapter 7)
✓ Employ an effective marketing mix (Chapter 8)
✓ Develop an attractive and user-friendly proprietary website (Chapter 9)
✓ Partner with appropriate third-party content-controlled sites (Chapter 10)
✓ Manage their reputations on UGC sites featuring reviews (Chapter 11)
✓ Regularly assess their marketing efforts (Chapter 12)

When foodservice operators focus on their guests and properly utilize the many marketing tools available to them, these operators' success in marketing will be assured.

What Would You Do? 12.2

"I don't know, Thalia," said Lenny. "It seems to me that if that's how our customers prefer to buy from us, then that's where we should spend most of our advertising dollars."

Lenny, the assistant manager of Damiano's Coney Island, was talking to Thalia Papadopoulos, the owner of Damiano's. Thalia was the restaurant's third-generation owner, and Lenny and Thalia were discussing the results of last year's marketing plan as they prepared for next year's plan.

"I agree that our takeout business is booming," said Thalia, "But since we serve breakfast, lunch, and dinner, I think we need to take a hard look at our breakfast business. That's been holding steady, but I think we can improve it with some targeted marketing for those guests who still want to take the time for a leisurely breakfast."

"Well," said Lenny, "We do get great reviews on our breakfast, and I think that's because of our Greek-style fried potatoes. Everybody raves about them."

"Exactly," replied Thalia, "I think if we promote that item specifically it could go a long way toward improving our breakfast sales."

"O.K.," said Lenny "But no one is ordering carryout breakfast. It's our Coney dogs at lunch that drove most of our sales increase last year."

"I do agree with that!" said Thalia "Thank goodness for Grandpa Leonidas' original coney dog sauce recipe!"

Assume you were Thalia. What factors should you consider as you decide whether next year's marketing efforts should be directed toward part of your business that is already extremely popular? Alternatively, are carryout Coney dogs an upward trend, or should an area with good market potential for increased popularity (dine-in breakfast) be increasingly emphasized? Explain your answer.

Key Terms

Revenue metric	Nonfinancial metric	Market trend
Marketing return on investment (ROI)	SWOT analysis	Market potential

---

Operator's 10-Point Tactics for Success Checklist

Evaluate your need for, and the current status of, each of the following operational tactics. For those tactics you think are important, but not yet in place, develop an action plan for its implementation including who will be responsible for the tactic's completion and the target date by which it should be completed.

Tactic	Don't Agree (Not Done)	Agree (Done)	Agree (Not Done)	If Not Done — Who Is Responsible?	If Not Done — Target Completion Date
1) Operator recognizes the importance of regularly evaluating marketing efforts.	____	____	____		
2) Operator has assessed the effectiveness of marketing strategies and tactics included in their most recently completed marketing plan.	____	____	____		
3) Operator understands the methods to evaluate changes in total revenue and guest counts.	____	____	____		
4) Operator understands the methods used to evaluate changes in guest check averages without menu price adjustments.	____	____	____		
5) Operator understands the methods used to evaluate changes in guest check averages with menu price adjustments.	____	____	____		

(Continued)

Tactic	Don't Agree (Not Done)	Agree (Done)	Agree (Not Done)	If Not Done	
				Who Is Responsible?	Target Completion Date
6) Operator understands the methods used to evaluate changes in revenue based on service style.	____	____	____		
7) Operator recognizes the importance of regularly monitoring the quality and impact of their website and online social media presence.	____	____	____		
8) Operator recognizes the importance of monitoring and managing guest rating scores appearing on Internet rating sites.	____	____	____		
9) Operator understands the importance of the SWOT analysis in the preparation of future marketing plans.	____	____	____		
10) Operator has the knowledge and tools required to create an effective marketing plan for the next marketing period.	____	____	____		

Glossary

4 Ps of Marketing: A means of categorizing a business' marketing strategy on the basis of the products sold, places where they are sold, the prices at which they are sold, and the promotional efforts used to sell them.

À la carte (menu): A menu that lists and prices each menu item separately.

Accuracy in menu laws: Legislation that requires foodservice operations to represent the quality, quantity, nutritional value, and price of the items they sell truthfully and accurately. Also known as "Truth-in-menu laws" and "Truth-in- dining laws."

Advertising campaign: A specifically designed strategy conducted across different communication channels to achieve specific results including increased brand awareness, higher sales, and improved communication with a specific target market.

Advertising: Any form of marketing message delivery for which a foodservice operation must pay.

AIDA: An advertising tool used as a guide to help ensure an advertisement is effective. AIDA is an acronym for attention, interest, desire, and action.

Arithmetic average: The simplest and most widely used measure of a mean (average); calculated by dividing the sum of a group of numbers by the count of the numbers used in the series.

Attraction: A place or event that draws visitors by providing something of interest or pleasure.

Back-of-house staff: The employees of a foodservice operation whose duties do not routinely put them in direct contact with guests.

Beer: An alcoholic beverage made from malted grain and flavored with hops.

Bounce rate: The percentage of visitors who view only one page before exiting a website.

Brand identifiers: The visible features that create a specific company brand. These can include any combination of name, logo, exterior building design, interior design, décor, music, menus, uniforms, pricing structure, services offered, and digital and traditional messaging.

Brand mark: The symbol or logo used to identify a foodservice operation. The "Golden Arches" used in McDonald's advertising are a good example of a foodservice company's well-known brand mark.

Brand: The specific ways in which a foodservice operation differentiates itself from its competitors.

Bundling: A pricing strategy that combines multiple menu items into a grouping which is then sold at a price lower than that of the bundled items purchased separately.

Bureau of Alcohol, Tobacco, and Firearms (ATF): The agency in the U.S. government responsible for, among other things, monitoring the manufacture and sale of alcoholic beverages.

Business plan: A document that summarizes the operational and financial objectives of a business.

Buyer's remorse: A feeling of regret that occurs after buying a product or service.

Buy-one-get-one (BOGO): A sales promotion in which an item is offered for free when another of the same item is purchased at full price.

Call to action: A piece of content intended to induce a viewer, reader, or listener to perform a specific act, typically taking the form of an instruction or directive.

Carafe: A container used for serving wines. A standard carafe holds one standard size (750 ml) bottle of wine. However, carafe size may vary based on an operation's own service preferences.

Cash flow: The net amount of cash and cash equivalents transferred in and out of a company. Cash received represents inflows, and funds spent represent outflows.

Chain (restaurant): A group of restaurants with many different locations that share the same name and concept. Chain restaurants may be owned and operated by the chain, and/or they may be individually owned through franchising. Chain restaurants may be classified as QSRs, fast casual, casual or up-scale operations.

Click-through rate (CTR): The number of clicks an ad or promotion receives divided by the number of times the ad is viewed. The formula used to calculate CTR is:

$$\frac{\text{Number of Clicks}}{\text{Number of Views}} = \text{CTR}$$

Community relations: The work an operation does to establish and maintain goodwill in its community.

Consumer rationality: The tendency to make buying decisions based on the belief that the decisions are of personal benefit.

Content management system (CMS): Computer software used to load content into a digital menu display system.

Contribution margin (CM): The amount of revenue that remains after a menu item's food cost is subtracted from its selling price.

Convenience food: Any food item that is partially, or fully, prepared before it is purchased by a foodservice operation.

Convention and Visitors Bureau (CVB): The local entity responsible for promoting travel and tourism to and within a specifically designated geographic area.

Corporate social responsibility (CSR): Those business actions designed to enhance society and the environment instead of contributing negatively to them.

Craft beer: A beer produced in limited quantities and with limited availability.

Crisis management plan (CMP): Written instructions that describe how an operation will react to a crisis, including who will be involved in the response and what will be done.

Cross-training: The action or practice of training workers in more than one role or skill.

Curb appeal: The general attractiveness of a foodservice operation when viewed from the outside by a potential customer.

Customer loyalty program: A marketing tool that recognizes and rewards customers who make purchases on a recurring basis. The programs typically award points, discounts, or other benefits that increase as the total amount of a repeat customer's purchases increase. Also commonly known as a "Frequent guest program," "Frequent customer program," or "Frequent dining program."

Cyclical (menu): A menu in which items are offered on a repeating (cyclical) basis.

Digital marketing metric: A key performance indicator (KPI) used to measure the success of a business's online marketing efforts. The goal of using digital marketing metrics is to track, record, and assess how consumers interact with a business online through websites and social media platforms.

Digital menu: An integrated system that uses hardware and software to display an operation's menu on an electronic screen; also commonly referred to as a digital display menu or digital menu board.

Discretionary income: The amount of disposable income remaining after all basic necessities such as housing, food, and clothing are paid.

Disposable income: The amount of income left over after all taxes are paid. Also known as net personal income.

Domain name: The unique name that appears after www. in web addresses and after the @ sign in e-mail addresses.

Draft beer: An unpasteurized beer product sold in kegs; also known as "tap" beer.

Du jour (menu): A menu featuring items and prices that change daily. Pronounced "duh-zhoor."

Empowerment (staff): An operating philosophy that emphasizes the importance of allowing staff to make independent guest service-related decisions and to act on them.

Ethics: The moral principles that govern a person or business organization's conduct toward others.

Extension (domain name): The combination of characters following the period in a web address.

Fake review: A positive, neutral, or negative review that is not an actual consumer's honest and impartial opinion or that does not reflect a consumer's genuine experience with a product, service, or business.

Farm-to-fork: A procurement process that stresses the importance of cooking with the freshest, locally, and sustainably grown seasonal ingredients. Also known as "farm-to-table."

Federal Trade Commission (FTC): The agency in the U.S. government responsible for protecting consumers and competition by preventing anticompetitive, deceptive, and unfair business practices.

Fine dining (restaurant): A restaurant experience that is typically more sophisticated, unique, and expensive than one would find in the average restaurant. Also known as an upper-scale restaurant.

Font size: The size of the characters printed on a page or displayed on a screen. Commonly referred to as type size.

Food and Drug Administration (FDA): The agency in the U.S. government responsible for assuring that foods (except for meat from livestock, poultry, and some egg products which are regulated by the USDA), are safe, wholesome, sanitary, and properly labeled.

Food cost percentage: A ratio calculated by determining the food cost for a menu item and dividing that cost by the selling price of the item.

Franchised operation: A business in which a foodservice operator (the franchisee) is allowed to use the brand identifiers of an organization (the franchisor) in exchange for the payment of franchise fees.

Frequency (advertising): The total number of times a specific viewer or listener is exposed to an advertisement.

Front-of-house staff: The employees of a foodservice operation whose duties routinely put them in direct contact with guests.

F-shaped Scanning pattern: A text scanning pattern characterized by fixations concentrated at the top and the left side of a page. Website viewers typically first read in a horizontal movement across the upper part of the content area. Note: This initially forms the F's top bar, and viewers then scan down. The "F" in F-shaped scanning pattern is short for "Fast," and the method is also commonly referred to as the "F-based reading pattern."

Ghost kitchen: A foodservice operation that provides no on-premises dine-in services and that prepares all of its menu items only for pick-up or delivery to guests. Also known as a delivery-only restaurant, shadow kitchen, cloud kitchen, or virtual kitchen.

Gig economy: A labor market characterized by the prevalence of short-term contracts or freelance work as opposed to more permanent jobs.

Google search: Also known simply as "Google;" this is a method of finding information on the Internet. Note: Google conducts over 92% of all Internet searches and is the most visited website in the world.

Grand opening: A well-publicized event that marks the first day of operations for a foodservice business.

Green practices: Those activities that lead to more environmentally friendly and ecologically responsible business decisions. Also known as "eco-friendly" or "earth-friendly" practices.

Guest check average: The average (mean) amount of money spent per guest (or table) during a specific accounting period. Also referred to as "check average" or "ticket average."

Hybrid delivery program: An arrangement in which a foodservice operator and a third-party delivery app jointly offer delivery services for the operator's store(s).

Income and expense statement: A detailed listing of revenue and expenses for a given accounting period. Also commonly referred to as an "Income Statement," "Profit and Loss Statement," or "P&L."

IP address: A unique address that identifies a device on the Internet or a local network. IP stands for "Internet Protocol," which is the set of rules governing the format of data sent via the Internet.

Keywords: The words and phrases that users type into search engines to find information about a particular topic. Also known as "SEO keywords," "Keyphrases," or "Search queries."

Link: A means of easily navigating from one webpage to another or from one section of a webpage to another section of the same webpage.

Local link: Website links created to show that other entities with relevance in the local area trust or endorse a foodservice operation.

Market potential: An estimate of total dollars or other units of measure for a specific product including those sold by an operation's competitors.

Market trend: Quantifiable changes in the behavior of a target market such as what target members buy, how they buy, and when they buy.

Marketing calendar: A schedule detailing all marketing activities planned for the time period covered by the marketing plan.

Marketing channel: A method or form of delivering a marketing message to an operation's current and potential guests. Also commonly known as a "communication channel."

Marketing message: How an operation communicates to its customers and highlights the value of its products and services.

Marketing mix: The specific ways a business utilizes the 4 Ps of marketing to communicate with its potential customers.

Marketing plan: A written plan detailing the marketing efforts of a foodservice operation for a specific time period (usually annually).

Marketing return on investment (ROI): The amount of benefit gained from a marketing activity compared to the initial cost of the activity.

Marketing strategies: The broad and long-term marketing goals a foodservice operation wishes to achieve.

Marketing tactics: The specific steps and actions undertaken to achieve a marketing strategy (goal).

Marketing: The varied activities and methods used to communicate a business' product and service offerings to its current and potential customers

Meeting planner: An individual responsible for securing the meeting space, food, and other services required by those attending group meetings. Also known as a Meetings Professional or Event Planner.

Menu copy: The words and phrases used to name and describe items listed on a foodservice menu.

Menu engineering: A system used to evaluate menu pricing and design to create a more profitable menu. It involves categorizing menu items into one of four categories based on the profitability and popularity of each item.

Menu: A French term meaning "detailed list." In common usage it refers to (i) a foodservice operation's available food and beverage products and (ii) how these items are made known to guests.

Merchant services provider: A company that enables foodservice operators (merchants) to accept credit and debit card payments and other alternative payment methods. Also sometimes referred to as a "payment services provider."

Mobile wallet: A virtual wallet that stores payment card information on a mobile device. Mobile wallets provide a convenient way for a user to make in-store payments and can be used at merchants listed with the mobile wallet service provider. Also known as a digital wallet.

Moment of truth: A point of interaction between a guest and a foodservice operation that enables the guest to form an impression about the business.

Navigation (website): The process of moving from one content section of a website to another section.

Net income: The profits achieved in a foodservice operation after income taxes have been paid.

Non-Commercial (foodservice operation): Foodservice in an organization where generating food and beverage profits is not the organization's primary purpose.

Nonfinancial metric: A standard for measuring or evaluating something, especially one that uses nondollar figures or statistics. Also commonly referred to as a "key performance index (KPI)."

Off-premise menu: A menu listing items available for pick up or delivery; also commonly referred to as a takeaway, take-out, or carry-out menu.

Online ordering system: A software program that allows a foodservice operation to accept and manage orders placed via a website or a mobile app.

Paid social (advertising): The use of ads on social media sites for which an operator must pay a fee.

PDF: Short for "Portable Document Format." A document format that allows documents to be shown clearly regardless of the software, hardware, or operating system utilized by the viewer.

Personal selling: Face-to-face selling-focused interactions that take place between foodservice personnel and their customers.

Point-of-sale (POS) system: An electronic system that records foodservice customer purchases and payments, as well as other operational data.

Portion cost: The product cost required to produce one serving of a menu item.

Preopening marketing expense: A nonrecurring promotional or advertising cost incurred before a new operation opens. Also referred to as a "start-up" marketing cost.

Press release: A communication tool used to objectively announce something newsworthy. Its purpose is to obtain media coverage and be noticed by an operation's target market.

Price (noun): A measure of the value given up (exchanged) by a buyer and a seller in a business transaction.

Price (verb): To establish the value to be given up (exchanged) by a buyer and a seller in a business transaction.

Prime (beef): The highest quality grade given to beef graded by the USDA. Prime beef is produced from young, well-fed beef cattle and has abundant marbling (the amount of fat interspersed with lean meat).

Prime real estate (menu): A phrase used to define the areas on a menu that are most visible to guests and, therefore, which should contain the items menu planners most want to sell: those that are most popular and profitable.

Profit margin: The amount by which revenue in a foodservice operation exceeds its operating and other costs.

Promotion (sales): A product or service offered for a limited time designed to encourage guests to make a purchase within that time.

Promotional code: A series of letters or numbers that allow a consumer to receive a discount on a specific purchase.

Proof (alcoholic beverage): A measure of the alcohol content of an alcoholic beverage. In the United States, alcohol proof is defined as twice the percentage of "alcohol content in percent by volume," or ABV.

Proprietary website: A website in which the foodservice operator controls all the website's content and can readily make changes to it.

Public relations (PR): The various ways a foodservice operation communicates with the media and the public to build mutually beneficial relationships; also commonly referred to as "PR."

Publicity: Free media attention for a product, service, business, or employee. It can include the use of traditional news sources such as televised news programs, magazines, and newspapers and through newer media such as podcasts, blogs, and websites.

QR code: A QR (quick response) code is a machine-readable bar code that, when read by the proper smart device, allows foodservice guests to view an online menu, be redirected to an online ordering website or app, or that allows them to order and/or pay for their meal without having to interact directly with a foodservice operation's staff.

Quick service restaurant (QSR): Foodservice operations that typically have limited menus, and most often include a counter at which customers can order and pick up their food. Most QSRs also have one or more drive-through lanes that allow customers to purchase menu items without leaving their vehicles. Menu prices in QSRs will be relatively low compared to other restaurant types.

Reach (advertising): The total number of viewers or listeners who could be exposed to an ad in a specific time frame.

Registration (domain name): The act of reserving a domain name on the Internet for a fixed period of time, usually one year. Registration is necessary because domain names cannot typically be purchased permanently.

Restaurant row: A street or region well-known for having multiple foodservice operations within close proximity.

Revenue metric: A standard for measuring or evaluating data based on its dollar amount or change in dollar amount. Also known as a "financial metric."

Sales call: A conversation between a salesperson and a prospective customer about the purchase of a product or service.

Search engine optimization (SEO): The process of improving and modifying an operation's website with the purpose of increasing its visibility when users search for products or services related to the operation.

Search engine results page (SERP): The webpage a search engine such as Google, Bing, or Yahoo shows a user when the user types in a search query.

Service industry: A business segment that primarily provides services to its customers.

Social media: A collective term for websites and applications that enable users to create and share information and content or to participate in social networking.

Sous vide: A method of cooking food that has been vacuum-sealed in a plastic bag and immersed in a regulated, low-temperature water bath.

Spirit (beverage): An alcoholic beverage produced by the distillation of fermented grains, fruits, vegetables, or sugar.

Spot (broadcast): A multi-media advertisement that is broadcast at a specific time.

Standardized recipe: The ingredients needed, and the procedures used to produce a specific menu item.

SWOT analysis: Strengths, Weaknesses, Opportunities, and Threats analysis; a systematic approach to assessing an operation's marketing strategies and tactics planning.

Table d'hôte (menu): A menu with a pre-selected number of menu items offered for one fixed price. Pronounced "tah-buhlz-doht." Also known as a "prix fixe" (pronounced prē-'fēks) menu.

Table management system: A software program, interfaced with a foodservice operation's POS system, that efficiently manages the operation's guest reservation and seating processes.

Target market: The group of people with one or more shared characteristics that an operation has identified as most likely customers for its products and services.

Third-party delivery: A smartphone or computer application that creates a marketplace that customers can search to browse restaurant menus, place orders, and have them delivered to a location of the customer's choosing. In nearly all cases, the guest orders are delivered by independent contractors who have been retained by the company operating the third-party delivery app.

Trademark: A brandmark that has been given legal protection and is restricted for exclusive use by the owner of the trademark.

Traffic tracking (website): Information that tells website operators how many site visitors it has had, how the site was found, how long the visitors browsed different pages on the website, and how frequently the visitors were converted into customers.

Uniform System of Accounts for Restaurants (USAR): A recommended and standardized (uniform) set of accounting procedures for categorizing and reporting restaurant revenue and expenses.

United States Department of Agriculture (USDA): The agency in the U.S. government responsible for regulating the production and sale of most food products manufactured and sold in the United States.

Up-selling: The process of increasing the guest check average through the effective use of on-premise selling techniques; also referred to as "suggestive selling."

User experience: The feelings resulting from a user's interactions with a digital product.

User-generated content (UGC) site: A website in which content including images, videos, text, and audio have been posted online by the site's visitors.

Value menu: A group of menu items designed to be the least expensive items available. The portion size and number of items included in a value menu differ by operation, but all are designed to attract consumers with low-priced menu items.

Value: The amount paid for a product or service compared to the buyer's view of what they receive in return.

Variance: The difference between an operation's actual and estimated income or expense.

Wagyu Beef: Beef from a Japanese breed of cattle that is highly prized for its marbling and flavor. In the Japanese language, "Wa" means Japanese, and "gyu" means cow.

Web host: An organization that provides space for holding and viewing the contents of a website. All websites and e-mail systems must be hosted to connect an operator's proprietary website content to the rest of the Internet.

Web-based communication channel: A communication method that relies on the internet and/or a smart device for its message delivery. Also referred to by some marketing experts as a "Digital" or "Online" communication channel.

Weighted contribution margin: The contribution margin provided by all menu items divided by the number of items sold. Weighted contribution margin is calculated as:

$$\frac{\text{Total Contribution Margin of All Items Sold}}{\text{Number of Items Sold}} = \text{Weighted contribution margin}$$

Wine list: A foodservice operation's wine menu.

Wine: An alcoholic beverage made from fruit; most typically grapes.

Index

a

Accuracy in menu laws 39
Advertising
  AIDA tool 77–78
  choosing the best channel 178–179
  defined 173–174
  effective 175
  frequency 178
  goal of 194
  message development process
    176–178
  paid social 248
  print 74–76
  promotion 9
  purpose of 174–176
  radio 76–77
  reach 178
  television 74, 77–78
  UGC sites 247–249
Advertising campaign 176
AIDA advertising tool 77–78
À la carte menu 147–148
Amazon Web Services 203
Appetizers 152, 155
Arithmetic average 260
Assessment
  financial 274–281
  marketing ROI 271

  marketing strategies 269–270
  marketing tactics 270–273
  nonfinancial 281–284
  nonfinancial metrics 273
  revenue metrics 268
  SWOT analysis 284–288
  tools 273–284
Attraction 63
Average cost of meals served 133

b

Baby Boomers 28
Background music 8
Back-of-house staff 10
Beer 157
Behavioral segmentation 27–28
Beverage menu 157
  beer menu 157–158
  spirit menu 159–160
  wine lists 158–159
Beverages (non-alcoholic) 153
Bounce rate 212, 282
Bourbon 160
Brand
  awareness 105
  defined 50
  developing unique 51–53
Brand identifiers 50

Branding
  defined 50
  marketing messages and 49–55
  product-focused 54
  service-focused 54–55
Brand mark 51
Broadcast communication channels 76
  radio 76–77
  spot 76, 77
  television 77–78
Buffet 54
Bundling 130
Bureau of Alcohol, Tobacco, and
       Firearms (ATF) 90
Burger King 58, 60
Business plan 102
Buyer behavior 86–87
  evaluation of purchase decision 88
  information assessment 87–88
  information gathering 87
  need recognition 87
  purchase decision 88
Buyer classification
  Friedman on 83–84
  #1 type 84
  #2 type 84–85
  #3 type 85–86
  #4 type 86
Buyer's remorse 255
Buy-one-get-one (BOGO) 184

C

Call to action 249
Carafe 158
Carryout 128
Cartwheel company 242
Cash flow 239
Casual service restaurants 67–68
Chain (restaurant) 41
Cheerful behavior 15
Chicken Bones restaurant 211
Cholesterol-free foods 151

Click-through rate (CTR) 212,
  282–283
Cloud-based POS systems 44
Cocktail menu. *See* Spirit menus
Coding coupons 272
Communication channel. *See* Marketing
  channels
Community relations 192–193
Competition 40, 125, 233
  similar prices 41–42
  similar products 40–41
  similar service style 41
Concession stands 54
Consistency in service 16
Consumer rationality 123
Content management system (CMS)
  162–163
Contribution margin (CM) 135.
  See also Weighted
  contribution margin
Contribution margin-based
  pricing 135–137
Convenience 61–62
Convenience food 150
Convention and Visitors Bureau (CVB)
  182–183, 225
Corporate social responsibilities
  (CSR) 91–92
Cost-based pricing 132–135
Cost-conscious customers 30
COVID-19 pandemic 28
Craft beer 158
Crisis management plan (CMP)
  190–191
Cross-training 36–37
Curb appeal 7
Customer attraction 204
Customer loyalty program 27, 185, 209
Customer relations management (CRM)
  programs 11
Customer's need 29–32, 87
Cyclical menu 148

d

Delivery
  method 12–13, 128–129
  take away items 168
Demographic segmentation 25–26
Desserts 153
Digital marketing metrics
  conversions 80
  defined 79
  engagement 80
  traffic 79–80
Digital menu
  content management system 163
  defined 162
  file size 165
  font size 164
  on-site 162
  on-user 163
  page size 163–164
Dine-in service 128
Discretionary income 39
Disposable income 39
Domain name
  defined 198
  extension 199
  finalizing 200
  high-quality 201
  hyphens and double letters 200
  pronounciation and spelling 200
  registration 200–201
  short and precise 199–200
Domino's Pizza 239–240
Draft beers 157
Drive-through 34–35, 62
Du jour menu 148

e

Effective marketing 3
E-mail marketing 106–107
Employee relations 193
Employee scheduling 37

Empowerment (staff) 38
Entrée category 152
Ethics 90–91
Extension (domain name) 199

f

Facebook 217, 229
Fake review 253
Farm-to-fork 43
Fast casual restaurants 66–67
Federal Trade Commission (FTC) 89
Fine dining (restaurant) 52, 68–69
Font size 164
Food and Drug Administration
  (FDA) 89
Foodborne illness 10, 39, 189
Food cost percentage 132
Foodies (food people) 281
Food menu development process 151
  design choices 155–156
  identification of menu categories
    152–153
  selection of menu items 152–153
  writing menu copy 154–155
Food photographs 230
Food safety, training tools 10
Food trucks 54
4 Ps of marketing
  defined 5
  illustration of 6
  place 7–8
  price 8–9
  product 6–7
  promotion 9–10
#4 type buyers 86
Franchised operation 52
Free to Choose 83
Frequency (advertising) 178
Friedman, Milton 83–84
Front-of-house staff 16
F-shaped scanning pattern 206–207

g

Generation Alpha   28
Generation X   28–29
Generation Z   28
Geographic segmentation
   24–25
Geo-targeting   63
Ghost kitchen   4
Ghost operations   69
Ghost restaurants   55
Gift cards   85, 209
Gig economy   37–38
Gluten-free foods   151
GoDaddy   203
Good sales records   31–32
Google Analytics   211, 212
Google Business Profile   229
Google Cloud   203
Google search   205
Grand opening   174–175
Green practices   26, 43
Groupthink   256
Guest check averages   152
Guest loyalty programs   27

h

Healthy foods   151
Heart-healthy foods   151
Hunger   29
Hybrid delivery program
   240–241

i

Incentives   272
Income and expense statement   236
Inconsistency   16
Information assessment   87–88
Information gathering   87
In-person selling   180
Inseparability   15–16
Instagram   218

Internet Corporation for Assigned
   Names and Numbers
   (ICANN)   201
IP address   198–199

j

James Beard Foundation Awards   57

k

Keywords   213–215

l

Leadership in Energy and
   Environmental Design (LEED)
   certification   130
Legislation   39–40, 89–90
Limited capacity   17–18
Links   223
Listening
   to customers   30
   to staff   36
Local links   223–226
Location   62–64, 129–130
Low-fat foods   151
Loyal customers   185–186

m

Marketing
   defined   3
   effective   3
   goal of   38
   importance of   2–5
Marketing budget   102–103, 108
Marketing calendars   103
Marketing channels   69, 73
   new channel assessment   82–83
   optimizing the frequency of
      delivery   178–179
   selection   103, 178–179
   traditional   74
      AIDA tool   77
      print   74–76

radio 76–77, 106
television 77–78
web-based 78–81
conversion 80
defined 74
digital marketing metrics 79–80
e-mail 106
engagement 80
traffic 79–80
UGC sites 79
Marketing communication
one-way 73
two-way 74
Marketing effort evaluation
financial 274–275
changes in revenue generation by
service method 279–281
changes in total revenue
generation 275
guest check averages 277–279
guest counts 275–277
importance of 267–273
marketing ROI 271
marketing strategies 269–270
marketing tactics 270–273
nonfinancial 281
rating scores on UGC sites
283–284
social media efforts 283
website traffic 282–283
revenue metrics 268
setting next-period strategies and
goals 288–289
SWOT analysis 284–288
Marketing message 28
choice of operation's name 53
delivery 73–83
effective 50
importance of branding 49–55
operator responsibilities 88
ethical 90–91
legal 89–90

social 91–92
product-related 55–64
service levels 64–70
Marketing mix 5, 174
advertising 173–179
personal selling 179–183
promotions 183–186
publicity 187–191
public relations 191–194
summary of 194
Marketing plan
advantages 98–99
challenges 99–100
cost estimates 108–109
defined 96
development 101
determining marketing budget
102–103
determining marketing goals
101–102
marketing calendars 103
marketing channel selection 103
target market identification 102
task assignment 103
evaluation 115–117
implementation 109–114
importance of staff efforts
111–112
delivery of guest orders 113
greeting guests 112
order-taking 112–113
payment collection 114
serving guests 113
purpose of 96
Marketing return on investment
(ROI) 271
Marketing strategy/goal 97, 101–102,
104–106, 107, 269–270
Marketing tactics 270–273
completion schedule 110
defined 97
identification of 106–108

Market potential 288
Market share 105
Market trends 288
McDonald's 25, 41, 51, 58, 60, 176, 182
Meal period 129
Meeting planners 86
Meeting Professionals International 86
Menu. *See also* Digital menu
  accuracy of information 150–151
  à la carte 147–148
  beverage 157–161
  cyclical 148
  defined 146
  development team 149
  du jour 148
  food 151–156
  legal aspects 149–151
  as marketing tool 147
  off-premise 165–169
  prime real estate 155, 156
  table d'hôte 148
  types of 147–148
Menu copy 57, 154–155
Menu engineering 137–138
  defined 137
  matrix 140–141
  popularity 138
  software 142
  weighted CM 138–141
Menu management programs 133
Menu modification 141–142
Menu pricing 124. *See also* Pricing
  bundling 130
  competition 125
  delivery method 128–129
  economic conditions 124–125
  guests, type of 126
  location 129–130
  meal period 129
  portion size 127–128
  product quality 126–127

service level 125–126
Merchant services provider 239
Millennials 28, 29
Mobile app 34
Mobile order(ing) 43–44
Mobile wallets 34
Moments of truth 111
Monitoring
  guest counts 272
  reviews 259–260, 263
  sales 272–273
Morris, Edna 59

n
Navigation (website) 207–208
Need recognition 29–32, 87
Negative publicity 189–191
Negative reviews 261–264
Net income 237
New communication channels 82–83
Non-commercial (foodservice
    operations) 65
Nonfinancial metrics 273

o
Off-premise menu 165–169
Off-premise personal selling 182
#1 type buyers 84
One-way marketing communication 73
Online ordering system 43–44,
    61–62, 208
Online promotions 210
Online reviews 10, 14, 44–45
On-premise personal selling 181
On-site digital menu 162
Operation's name, choice of 53
Operator direct delivery 128
Operator responsibilities 88–93
  ethical 90–91
  legal 89–90
  social 91–92

Operators' challenges
  in marketing products   11–14
    delivery method   12–13
    quality   11–12
    quantity   12
    value perception   13–14
  in marketing services   14–18
    consistency   16
    inseparability   15–16
    intangibility   15
    limited capacity   17–18

p

Packaging
  takeout   166–169
  tamperproof   169
Paid social (advertising)   248
PDF   202
Percentage variance   115–116
Personal selling   9, 194
  defined   179
  goals of   180
  off-premise   182
  on-premise   181
Physical facilities   33–35
Pick-up services   34, 62, 128
Pinterest   218
Place   7–8
Point-of-sale (POS) system   27,
    186, 275
Popeyes Louisiana Kitchen
    (Popeyes)   42
Popularity (number sold)   138
Portion costs   132
Portion size   127–128
Positive publicity   187–189
Preopening marketing expense   102
Press release   192
Price   59–61
  competition   41–42
  defined   121

4 Ps of marketing   8–9
  guest's view of   122–124
  menu design   150
  operator's view of   122–123
  revenue and   124
  selling   134
  and value   121
Pricing
  CM method   135–137
  cost-based   132–135
  evaluation   137–143
  factor   134–135
  food and beverage   131–137
  importance of   121
  menu   124
    bundling   130
    competition   125
    delivery method   128–129
    economic conditions
      124–125
    guests, type of   126
    location   129–130
    meal period   129
    portion size   127–128
    product quality   126–127
    service level   125–126
  for profits   120–124
Prime (beef)   150
Prime real estate (menu)   155, 156
Printed communication   74–76
Product(s)
  availability   40
  in 4 Ps of marketing   6–7
  quality   126–127
Product-focused branding   54
Product marketing, operators' challenges
    in   11–14
  delivery method   12–13
  quality   11–12
  quantity   12
  value perception   13–14

Product-related marketing
   messages   55–56
   convenience   61–62
   location   62–64
   preparation method   58
   price   59–61
   quality-focused   56–58
   serving size   58–59
Professional Rights Organization
   (PRO)   8
Profit margin   122
Promotion (sales)   9–10
   for all   184
   coupon-based   183
   defined   183
   for loyal customers   185–186
   online   210
   purpose of   183–184
Promotional code   80
Proof (alcoholic beverage)   159
Proprietary website
   204–212
   click-through rates   212
   competitive advantage   205
   content   208–211
   credibility   204–205
   customer attraction   204
   defined   204
   design
      appearance   206
      F-shaped scanning pattern
         206–207
      navigation   207–208
   goals of   204–205
   Google search   205
   local linkage   223–226
   NAP information   204
   search engine optimization
      212–215
   traffic tracking   211–212
Psychographic segmentation   26

Publicity   9–10, 194
   defined   187
   negative   189–191
   positive   187–189
Public relations (PR)   194
   community relations   192–193
   defined   191
   employee relations   193
   media affiliations   191–192
Purchase decision   88

q
QR code   44
Quality
   product   11–12, 56–58, 126–127
   service   17
Quantity   12
Quick service restaurants (QSRs)   13, 66

r
Radio advertisements   74, 76–77, 106
   live read   77
   sponsorship   77
   straight read   77
Reach (advertising)   178
Red Lobster   59
Registration (domain name)   200–201
Restaurant row   129
Revenue metrics   268
Revenue variance   274
Reviews   210
   fake   253
   Google   258
   monitoring   259–260, 263
   motivation of reviewers   252–255
   negative   261–264
   online   10, 14, 44–45
   response to   261
   from satisfied customers   260
   scoring system   259–261
Rude behavior   15

S

Salads   152
Sales calls   182
Search engine   213–214
Search engine optimization (SEO)
    212–215, 226
Search engine results page (SERP)
    213–214
Self-delivery   239–243
Self-service kiosks   44
Selling price   134
Service-focused branding   54–55,
    64–70
Service industry   14
Service marketing, operators' challenges
    in   14–18
  consistency   16
  inseparability   15–16
  intangibility   15
  limited capacity   17–18
Service quality   17
Serving size   58–59
Social media sites   257
  advertising   179
  call to action   249
  defined   216
  evaluation of operation's efforts   283
  Facebook   217, 229
  importance of   216–217
  increasing activity on   105
  Instagram   218
  paid social advertising   248
  Pinterest   218
  posting content on   217–219
  TikTok   218, 219
  Twitter   218
  YouTube   217–218
Social responsibilities   91–92
Soups   152
Sous vide   58
Special events   188

Spirit (beverage)   159
Spirit menus   159–161
Sponsorships   224–225
Spot (broadcast)   76, 77
Staff(s)
  attitude   15–16, 35
  capabilities   35–38
  cross-training   36–37
  empowerment   38
  fair wages and benefits   36
  gig economy   37–38
  leading by example   36
  listening to   36
  scheduling process   37
  training   36
Standardized recipe   132
Starbucks   43, 176
Sugar-free foods   151
SWOT analysis   284–288

t

Table d'hôte menu   148
Table management system   33
Table reservations   209
Take away
  menu item selection   166
  packaging   166–167
  pick up/delivery   168
Tamperproof packaging   169
Target markets
  definition   22
  external factors
    competitive environment
      40–42
    economic environment
      38–39
    legal requirements   39–40
    social environment   42–43
    technology   43–45
    vendor and product availability   40
  foodservice operator's unique   23

Target markets (*continued*)
  identification   102, 177
    behavioral   27–28
    demographic   25–26
    generation-related   28–29
    geographic   24–25
    methods   24–28
    need for   23–24
    psychographic   26
  identification of needs and desires
      29–32, 177
    good sales records   31–32
    listening   30
    questioning   30–31
  internal factors
    physical facility   33–35
    staff capabilities   35–38
Task assignment   103
Technology   43
  cloud-based POS systems   44
  mobile ordering   43–44
  new payment options   44
  online reviews   44–45
  self-service kiosks   44
Television advertising   74, 77–78
Third-party delivery   230–231. *See also*
      Self-delivery
  advantages of   232–234
  competitive pressure   233
  during COVID pandemic
      128–129, 231
  defined   4
  disadvantages of   234–239
  financial concerns   236–239
  guest ownership   236
  impact on profits   233–234
  incentive programs   235
  independent contractors   235
  popularity   232–233
  quality control   234–235
  tamperproof packaging   168–169

Third-party-operated websites
      226–231, 242
Threat identification, in SWOT
      analysis   285–287
#3 type buyers   85–86
TikTok   218, 219, 247
Trademark   51
Traffic tracking   211–212
Training (staff)   36
TripAdvisor   255
Twitter   218
#2 type buyers   84–85
Two-way marketing communication   74

u
Uniform System of Accounts for
      Restaurants (USAR)   236
United States Department of Agriculture
      (USDA)   90
Upscale restaurants   68–69
Up-selling strategy   181
User experience   207
User-generated content (UGC)
    sites   79
  advertising on   247–249
  guest reviews
    evaluation of   283–284
    fake review   253
    Google   258
    historical perspective   250–252
    increasing the number of
        256–258
    monitoring   259–260, 263
    motivation of reviewers   252–255
    negative   261–264
    popular sites   250
    response to   261
    reward   258
    from satisfied customers   260
    scoring system   259–261
  importance of   246–247

V

Value   84
 defined   13, 59
 perception   13–14
 price and   121
Value menu   60–61
Variance   115. *See also* Percentage
  variance
Vegetable accompaniments   152
Vendors, choice of   40

W

Wagyu Beef   127
Waitstaff attitude   15–16
Web-based communication
    channel   78–81
 defined   74
 digital marketing metrics   79
  conversion   80
  engagement   80
  traffic   79–80
 e-mail   106–107
 UGC sites   79
Web-based marketing
 domain name selection   198–201
 IP address   198–199
 link   223
 proprietary website   204–212

search engine optimization   212–215
 social media sites   216–220
 web host   201–203
Web host   201–203
Website(s). *See also* Proprietary website
 average time on page   282
 bounce rate   282
 branding   51
 click-through rate   282–283
 consumer reviews on   10, 14, 44–45
 number of visitors   282
 third-party-operated   226–231, 242
 traffic   79–80, 282–283
 visitors   105
Web traffic   211–212
Weighted contribution margin   138–141
Whiskey   160
Wine   158
Wine lists   158–159
Work schedules   37

Y

Year-over-year (YOY) comparison   107
Yelp   255, 259
YouTube   217–218, 247

Z

Zagat Survey   251–252